Stefan Zander

Preparation and Characterization of Cu/ZnO Catalysts

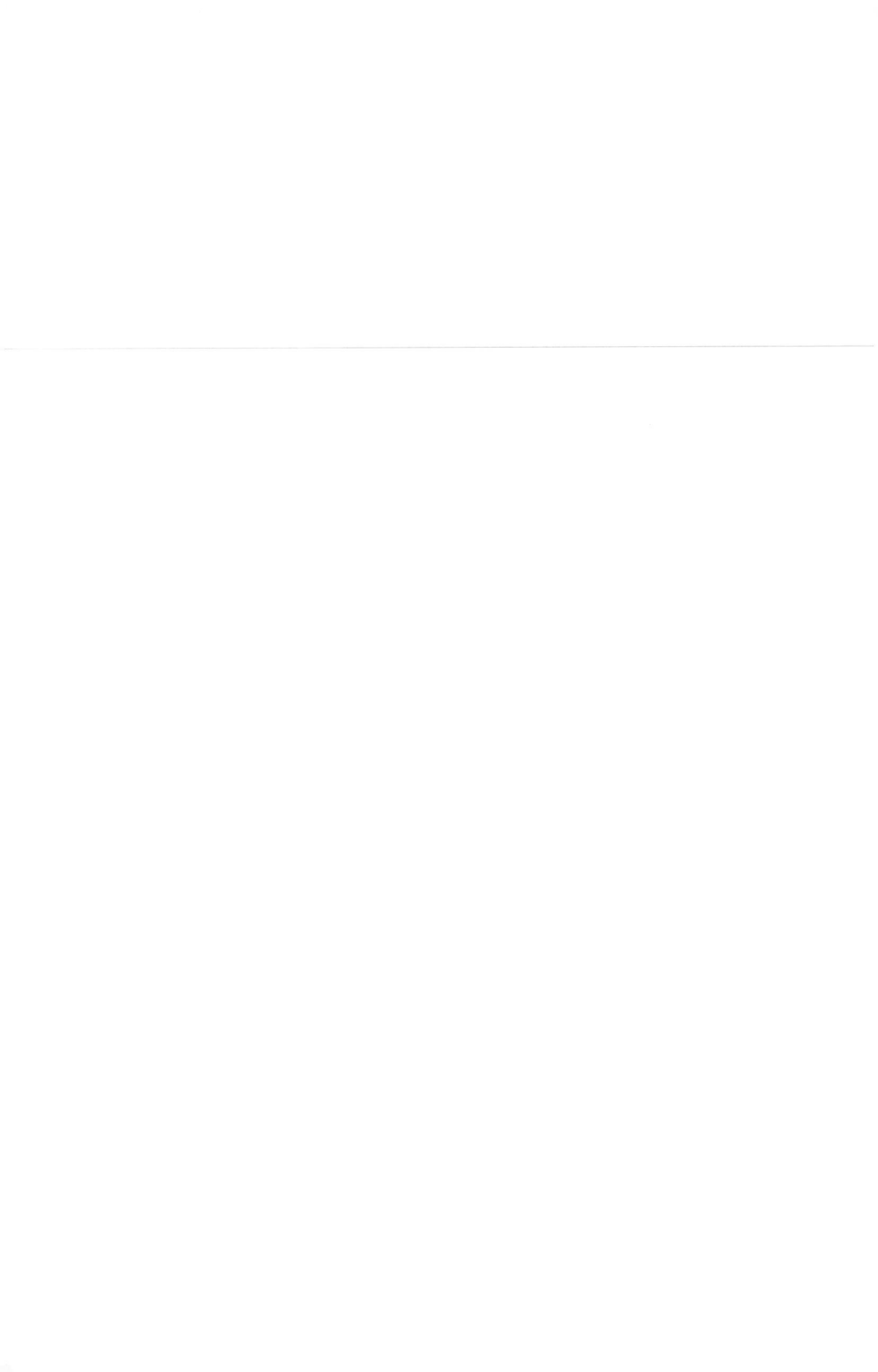

Stefan Zander

Preparation and Characterization of Cu/ZnO Catalysts

And their Application in Methanol Synthesis

Südwestdeutscher Verlag für Hochschulschriften

Impressum / Imprint
Bibliografische Information der Deutschen Nationalbibliothek: Die Deutsche Nationalbibliothek verzeichnet diese Publikation in der Deutschen Nationalbibliografie; detaillierte bibliografische Daten sind im Internet über http://dnb.d-nb.de abrufbar.
Alle in diesem Buch genannten Marken und Produktnamen unterliegen warenzeichen-, marken- oder patentrechtlichem Schutz bzw. sind Warenzeichen oder eingetragene Warenzeichen der jeweiligen Inhaber. Die Wiedergabe von Marken, Produktnamen, Gebrauchsnamen, Handelsnamen, Warenbezeichnungen u.s.w. in diesem Werk berechtigt auch ohne besondere Kennzeichnung nicht zu der Annahme, dass solche Namen im Sinne der Warenzeichen- und Markenschutzgesetzgebung als frei zu betrachten wären und daher von jedermann benutzt werden dürften.

Bibliographic information published by the Deutsche Nationalbibliothek: The Deutsche Nationalbibliothek lists this publication in the Deutsche Nationalbibliografie; detailed bibliographic data are available in the Internet at http://dnb.d-nb.de.
Any brand names and product names mentioned in this book are subject to trademark, brand or patent protection and are trademarks or registered trademarks of their respective holders. The use of brand names, product names, common names, trade names, product descriptions etc. even without a particular marking in this work is in no way to be construed to mean that such names may be regarded as unrestricted in respect of trademark and brand protection legislation and could thus be used by anyone.

Coverbild / Cover image: www.ingimage.com

Verlag / Publisher:
Südwestdeutscher Verlag für Hochschulschriften
ist ein Imprint der / is a trademark of
OmniScriptum GmbH & Co. KG
Heinrich-Böcking-Str. 6-8, 66121 Saarbrücken, Deutschland / Germany
Email: info@svh-verlag.de

Herstellung: siehe letzte Seite /
Printed at: see last page
ISBN: 978-3-8381-5020-8

Zugl. / Approved by: Berlin, Technische Universität Berlin, Dissertation, 2012

Copyright © 2014 OmniScriptum GmbH & Co. KG
Alle Rechte vorbehalten. / All rights reserved. Saarbrücken 2014

Abstract

In this work, systematic investigations of the preparation of Cu/ZnO-based methanol synthesis catalysts are presented. The catalyst precursors were prepared by co-precipitation, followed by aging, filtrating, washing and drying according to the proven, but not completely understood industrial preparation method. Subsequent calcination and reduction led to the catalytically active Cu/ZnO/X catalyst, X being a refractory oxide acting as structural promoter. The investigations focus on the chemistry of the zincian malachite precursor that was identified as the material yielding the best catalytic performance with the aim of identifying the role of the different synthesis parameters on its formation mechanism and properties and of establishing structure-performance-relationships that explain the role of the synthesis conditions and the structural promoter phase X on the final catalytic activity.

Co-precipitation (Cu:Zn = 70:30) was performed in a pH- and temperature-controlled (338 K) manner and enabled homogeneous distribution of the metal ions in the amorphous initial precipitate which transformed into crystalline zincian malachite during aging. This aging step was found to be critical with regard to the incorporation of Zn into zincian malachite and was investigated by *in-situ* methods. Therefore, it had to be decoupled from the prior co-precipitation step using co-precipitation with continuous spray-drying. As a function of aging pH, two different aging mechanisms were found that explain the effect of the synthesis conditions in the early stages of preparation on the structural properties of the precursor and later the resulting catalyst. Low pH-values (5.0-6.5) trigger a direct co-condensation mechanism, while at high pH values (7.0-8.0) a transient sodium zinc carbonate phase was observed upon crystallization of the zincian malachite precursor phase. The Zn incorporation into the zincian malachite precursor phase was higher at low pH values. Temperature was found to accelerate both pathways at a given pH value. Based on these results, the setting of the synthesis parameters in the applied catalyst preparation method can be rationalized. They have been optimized to yield maximal Cu,Zn substitution in zincian malachite which in turn is a precondition for final nanostructuring of the catalyst.

Also in conventional batch synthesis of Cu/ZnO catalysts the application of different pH-values in the range of pH 6.0-9.0 during co-precipitation was observed to influence the precursor chemistry. Application of pH values ≥ 6.5 led to higher phase fraction of zincian malachite at the expense of the undesired Zn-rich by-phase aurichalcite. As a consequence, more Zn was inserted into zincian malachite after aging, leading to smaller CuO domain size in the calcined catalyst. For pH-values in the basic regime, formation of two clearly different substituted zincian malachite phases was found indicating inhomogeneous Zn distribution in the precursor material. The pH-dependent switch of the aging mechanism observed during the previously described *in-situ* experiments is a likely explanation for the differences in Zn incorporation. However, the highest Cu surface area, which is a prerequisite for an efficient catalyst, was obtained for catalysts prepared at pH 8.5. Unfortunately, we were not able to track back this observation directly to the synthesis pH in a simple synthesis parameter–structure–performance relationship. The batch process is probably more complex as variation of the parameter pH may induce numerous changes in the precursor material that can lead to different and partially compensating effects for the resulting catalyst.

ZnO is known to act as a spacer for the single Cu particles in the Cu/ZnO catalyst and to enable the widely studied Cu-ZnO synergy which beneficially affects the activity. MgO was investigated to act as a substitute for ZnO following the substituted malachite preparation approach. At the same Cu content (80 mol%), the geometric influence turned out to be even better compared to ZnO but the synergetic effect of Cu and ZnO during methanol synthesis from $CO_2/CO/H_2$ was lacking. By subsequent impregnation with ZnO both geometric and synergetic effects were combined in a Cu/MgO/ZnO catalyst which exhibited a higher activity than Cu/ZnO and Cu/MgO. Thus, the geometric and synergetic effects of the oxide components have been separated during synthesis. Interestingly, if the feed gas was changed to CO/H_2, Cu/MgO was by far most active.

The effect of Ga_2O_3 as a promoter in the $Cu/ZnO/(Ga_2O_3)$ system was investigated by preparing a sample series with increasing Ga concentration. Ga contents up to 3 mol% were incorporated in the zincian malachite precursor despite the charge mismatch and changed the characteristics of the sample dramatically. After calcination, some of the Ga was incorporated in the ZnO. After reduction, the Cu surface area was increased by 100% and the methanol synthesis activity by 60% compared to the binary Cu/ZnO reference system. Higher Ga contents led to segregation and inhomogeneous microstructure of the resulting catalyst. The functionality of Ga promotion was found to critically depend on the homogeneous distribution of Ga. The best distribution was achieved by incorporation into the zincian malachite precursor phase and a linear correlation of the (Zn,Ga) content in this phase with the catalytic activity of the final catalyst was observed.

Zusammenfassung

Systematische Untersuchungen der Präparation von Cu/ZnO-basierten Methanolsynthese-Katalysatoren sind Gegenstand dieser Arbeit. Die Katalysatorpräkursoren wurden gemäß dem etablierten aber nur unzureichend verstandenen industriellen Syntheseweg hergestellt, der aus den Schritten Co-Fällung, Altern, Filtern, Waschen und Trocknen besteht. Calcinierung und Reduktion führen schließlich zum aktiven Cu/ZnO/X Katalysator, wobei X ein temperaturbeständiges Oxid darstellt, welches als struktureller Promotor fungiert. Die Untersuchungen richten sich auf die Chemie des Zink-Malachit-Präkursors, der letztendlich zu einer hohen katalytischen Aktivität führt. Dabei sollen die Einflüsse verschiedener Syntheseparameter auf die Bildung und Eigenschaften des Zink-Malachits und Mikrostruktur-Aktivitäts-Korrelationen untersucht werden, um die katalytische Aktivität mittels Syntheseparameter und Promotorphase X erklären zu können.

Durch pH- und temperaturkontrollierte (338 K) Co-Fällung (Cu:Zn 70:30) wird eine homogene Verteilung der Metallionen im anfänglich amorphen Fällungsprodukt erreicht, welches durch Altern zu Zink-Malachit kristallisiert. Die Alterung wird als entscheidender Schritt für die Einlagerung von Zink-Ionen im Zink-Malachit angesehen und wurde mit Hilfe von *in-situ* Methoden untersucht. Dafür war eine Entkopplung von der vorausgehenden Co-Fällung nötig, was durch Co-Fällung und kontinuierliche Sprühtrocknung realisiert wurde. Zwei verschiedene Alterungsmechanismen wurden, abhängig vom pH-Wert, beobachtet, welche den Einfluss der Syntheseparameter in den frühen Präparationsschritten auf strukturelle Eigenschaften des Präkursors und letztlich des Katalysators erklären können. Kleine pH-Werte (5.0-6.5) führen zu direkter Co-Kondensation, während für höhere pH-Werte (7.0-8.0) eine vorübergehende Natrium-Zink-Carbonat-Phase beobachtet wurde, bevor die Kristallisation der Zink-Malachit-Phase einsetzte. Bei kleinen pH-Werten konnte eine erhöhte Substitution von Cu-Ionen durch Zn-Ionen im Zink-Malachit erreicht werden. Erhöhung der Temperatur führte bei gegebenem pH-Wert zu einer Beschleunigung beider Mechanismen. Basierend auf diesen Erkenntnissen kann die Einstellung der Parameter bei dem vorliegenden Präparationsprozess vorgenommen werden. Ziel dabei ist eine möglichst große Einlagerung von Zink im Zink-Malachit, was wiederum eine Vorbedingung für die spätere Nano-Strukturierung ist.

Der pH-Wert spielt auch während der Co-Fällung eine entscheidende Rolle für die Eigenschaften des Cu,Zn-Präkursors (70:30) und wurde in zwei Serien von Batch-Synthesen im Bereich von 6.0-9.0 variiert. Wurden pH-Werte ≥ 6.5 verwendet, konnte nach dem Altern ein hoher Phasenanteil von Zink-Malachit neben der unerwünschten Zink-reichen Nebenphase

Zusammenfassung

Aurichalcit erzielt werden. Demzufolge konnte mehr Zink in Zink-Malachit eingebaut werden, was zu einer kleineren Kristallitgröße von CuO nach der Calcinierung führte. Für pH-Werte im basischen Bereich (≥ 7.5) wurden zwei deutlich verschieden substituierte Zink-Malachit-Phasen gebildet, die Zink-Verteilung im Präkursor war demnach inhomogen. Eine Erklärung für den Unterschied der Zink-Substitution liefert möglicherweise das Ergebnis des *in-situ* Alterungs-Experiments, wonach abhängig vom pH-Wert zwei verschiedene Mechanismen ablaufen können. Eine Voraussetzung für effektive Katalysatoren ist letztendlich die Cu-Oberfläche, die für bei pH 8.5 präparierte Proben am größten war. Leider konnte dieses Ergebnis nicht mit Hilfe einer Parameter-Struktur-Aktivitäts-Korrelation beschrieben werden. Der gesamte Batch-Prozess scheint daher sehr komplex zu sein, da die Variation des Parameters pH-Wert weitere Veränderungen im Präkursor hervorrufen kann, die zu verschiedenen und möglicherweise kompensierenden Effekten für den resultierenden Katalysator führen.

ZnO fungiert als Abstandshalter für einzelne Cu-Partikel in Cu/ZnO Katalysatoren und ermöglicht außerdem die Cu-ZnO-Synergie, welche die Aktivität positiv beeinflusst. Die Verwendung von MgO anstelle von ZnO wurde untersucht, wobei die Präparation der Cu/MgO-Katalysatoren analog zu der von Cu/ZnO erfolgte. Bei gleichem Cu-Gehalt (80 mol%) wurde für MgO ein besserer geometrischer Einfluss festgestellt, jedoch war kein vergleichbarer Effekt der Synergie während der Methanolsynthese aus $CO_2/CO/H_2$ messbar. Nachfolgende Imprägnierung mit ZnO führte zu einer Kombination der geometrischen und synergetischen Effekte in Gestalt eines Cu/MgO/ZnO Katalysators, der aktiver war als Cu/ZnO und Cu/MgO. Daher können geometrischer und synergetischer Effekt der Oxidkomponenten sequentiell während der Synthese eingeführt werden. Interessanterweise war Cu/MgO bei der Methanolsynthese aus CO/H_2 mit Abstand am aktivsten.

Der Effekt von Ga_2O_3 als Promotor im Cu/ZnO/(Ga_2O_3)-System wurde durch eine Probenserie mit ansteigender Ga-Konzentration untersucht. Bis zu 3 mol% Ga^{3+} konnten trotz der abweichenden Ladung in den Zink-Malachit-Präkursor eingebaut werden und führten zu einer drastischen Änderung der Probeneigenschaften. Nach der Calcinierung konnte ein Teil des Ga im ZnO nachgewiesen werden. In den reduzierten Proben wurde, verglichen mit der Cu/ZnO-Referenzprobe, die Cu-Oberfläche um 100% und die Aktivität in der Methanolsynthese um 60% erhöht. Höhere Ga-Gehalte führten zu Segregation und inhomogener Mikrostruktur des resultierenden Katalysators. Die beste Elementverteilung wurde erzielt, wenn Ga vollständig im Zink-Malachit-Präkursor eingebaut war. Es wurde eine lineare Korrelation zwischen dem (Zn,Ga)-Gehalt in Zink-Malachit und der katalytischen Aktivität gefunden.

Danksagung

Bedanken möchte ich mich besonders bei Herrn Prof. Dr. Robert Schlögl für die Möglichkeit, die vorliegende Arbeit in der Abteilung Anorganische Chemie des Fritz-Haber-Instituts der Max-Planck-Gesellschaft anfertigen zu können. Dabei bedanke ich mich für die interessante wissenschaftliche Fragestellung, die wertvollen Anregungen sowie für das sehr gute Arbeitsklima und die Arbeitsbedingungen in der Abteilung.

Mein Dank gilt ebenfalls meinem Gruppenleiter Dr. Malte Behrens für sein Vertrauen, die ausgezeichnete fachliche Betreuung, seinen diplomatischen und motivierenden Führungsstil und seine stete Diskussionsbereitschaft zu allen Aspekten der Arbeit.

Ich danke allen, auch den nicht namentlich genannten Mitarbeitern, die auf ihre Art und Weise zum Gelingen beigetragen haben, sei es durch praktische Hilfe oder fachliche Diskussionen. Der "Nanostructure"-Gruppe danke ich für die wissenschaftliche Unterstützung. Meiner ehemaligen Bürokollegin Dr. Antje Ota danke ich besonders für die gemeinsame Zeit mit vielen interessanten Diskussionen, vor allem abseits der Forschung, Dr. Thomas Cotter, Dr. Lénárd-István Csepei und Patrick Kast für das lockere Büroklima. Stefanie Kühl danke ich für die praktische Hilfe, vor allem in der ersten Zeit, und die unzählig vielen guten Tipps. Julia Schumann sei gedankt für die Unterstützung bei der Katalysatorpräparation, Dr. Edward Kunkes und Nygil Thomas für die katalytischen Messungen. Ich danke Dr. Frank Girgsdies, der immer ausführlich und klar auf meine Fragen antwortete, Dr. Igor Kasatkin für TEM-Messungen und die Diskussionen, Dr. Olaf Timpe für seine stete spontane Bereitschaft, mir bei vielen kleineren Problemen zu helfen. Ich danke den Personen am FHI für die Messung meiner Proben: Edith Kitzelmann (XRD und TG), Dr. Andrey Tarasov (TG), Gisela Lorenz und Maike Hashagen (BET), Gisela Weinberg (SEM), Jutta Kröhnert (IR), Genka Tzolova-Müller (UV-Vis), Doreen Steffen (Präparation), Achim Klein-Hoffmann (XRF) und Dr. Manfred E. Schuster (TEM).

Ich danke den Kooperationspartnern, Prof. Dr. Thorsten Ressler und Gregor Koch, TU Berlin, für XANES-Messungen, Dr. Wolfgang Bensch, Beatrix Seidlhofer, Elena Antonova und Jing Wang, Universität Kiel, für EDXRD-Messungen und dem HASYLAB (DESY), Hamburg, für die Bereitstellung von Strahlzeit.

Mein Dank gilt dem BMBF und dem STMWFK für die finanzielle Unterstützung im Rahmen der jeweiligen Projekte (01RI0529, NW-0810-0002), allen beteiligten Personen sowie den Mitarbeitern der Clariant Produkte (Deutschland) GmbH, Dr. Patrick Kurr, Dr. Benjamin Kniep und Dr. Nikolas Jacobsen.

Meiner Frau Meike und meinem Sohn Mateo danke ich für die Liebe und Kraft, die sie mir während dieser Zeit gegeben haben. Meinen Eltern, meinen Schwestern und meiner gesamten Familie danke ich für die große Unterstützung und Geduld während der Arbeit.

Table of Contents

Abstract i
Zusammenfassung iii
Danksagung v

List of Abbreviations x
Chapter 1: Introduction and Overview 1

1.1	Catalysis	1
1.2	Methanol synthesis over Cu/ZnO/Al$_2$O$_3$ catalysts	2
1.2.1	Methanol	2
1.2.2	Methanol synthesis	3
1.2.3	Preparation and Characteristics of Cu,Zn based catalysts	4
1.3	Outline of the work	9
1.4	References	10

Chapter 2: In-situ EDXRD Study of the Chemistry of Aging of Co-precipitated Mixed Cu,Zn Hydroxycarbonates – Consequences for the Preparation of Cu/ZnO Catalysts 13

2.1	Introduction	14
2.2	Experimental	17
2.2.1	Precursor Preparation	17
2.2.2	In-situ EDXRD and UV-Vis Spectroscopy during Simulated Aging	17
2.2.3	Ex-situ Characterization	18
2.3	Results and Discussion	20
2.3.1	General	20
2.3.2	Phase Evolution	20
2.3.3	The Effect of Temperature	27
2.3.4	The Effect of Acidity	30
2.3.5	The Effect of Potassium Counter Ions	31
2.4	Conclusion	33
2.5	References	35

Chapter 3: Correlations between Preparation and Microstructure of Cu/ZnO Catalysts for Methanol Synthesis – Influence of the pH value during Synthesis of Cu,Zn Hydroxy Carbonates   41

3.1	Introduction	42
3.2	Experimental	43
3.2.1	Sample Preparation	43
3.2.2	Characterization	44
3.3	Results and Discussion	46
3.3.1	Precipitation and aging	46
3.3.2	Precursor and calcined materials	49
3.3.3	Reduction and reduced samples	56
3.4	Conclusions	58
3.5	References	60
	Supplementary Information	61

Chapter 4: The Role of the Oxide Component in the Development of Copper Composite Catalysts for Methanol Synthesis   67

4.1	Introduction	68
4.2	Experimental	70
4.2.1	Catalyst Preparation	70
4.2.2	Characterization	71
4.2.3	Catalytic performance	71
4.3	Results and discussion	72
4.4	Conclusion	78
4.5	References	79
	Supplementary Information	80

Chapter 5: Promoting Methanol Synthesis Catalysts: Correlations between Microstructure and Activity in Cu/ZnO/Ga$_2$O$_3$ 81

5.1	Introduction	82
5.2	Experimental	83

5.2.1	Sample Preparation	83
5.2.2	Sample Labeling	84
5.2.3	Elemental Analysis	84
5.2.4	Characterization	84
5.2.5	Catalytic testing	86
5.3	Results and Discussion	87
5.3.1	The influence of gallia on the precursor chemistry	87
5.3.1.1	XRD analysis	87
5.3.1.2	Scanning electron microscopy	92
5.3.1.3	Thermal analysis	94
5.3.2	Calcined samples	97
5.3.2.1	XRD analysis	97
5.3.2.2	The influence of gallia on ZnO	98
5.3.2.3	Temperature programmed reduction	101
5.3.3	Activated samples	102
5.3.3.1	Transmission electron microscopy	102
5.3.3.2	Cu surface areas	104
5.3.4	Methanol synthesis activity	106
5.4	Conclusions	107
5.5	References	109

Chapter 6:   Final Summary and Conclusion   111

List of Abbreviations

BET	Adsorption isotherm model of Brunauer, Emmet and Teller
EDX	Energy dispersive X-ray spectroscopy
EDXRD	Energy dispersive X-ray diffraction
HRTEM	High resolution transmission electron microscopy
MS	Mass spectrometry
PSD	Particle size distribution
RFC	Reactive frontal chromatography
SA	Surface area
SEM	Scanning electron microscopy
TCD	Thermal conductivity detector
TEM	Transmission electron microscopy
TG	Thermo gravimetry
TPR	Temperature programmed reduction
UV-Vis	Ultraviolet-visible spectroscopy
XANES	X-ray absorption near edge spectroscopy
XRD	X-ray diffraction
XRF	X-ray fluorescence

Chapter 1: Introduction and Overview

1.1 Catalysis

Catalysis is one of the central concepts in chemistry for organic as well as inorganic processes. Catalysts accelerate certain chemical reactions by decreasing the activation energy of single elementary reaction steps. As a consequence, a reaction can be steered into a desired direction with the constraint that only the kinetics can be changed but not the thermodynamics.

Nowadays, around 90% of the chemical processes use heterogeneous catalysts in chemical, food, pharmaceutical, automobile and petrochemical industries. More modern fields are fuel cells, green chemistry, nanotechnology and biotechnology. The main advantage of heterogeneous catalysis is that the catalyst and the reaction products can be easily separated from each other. Biocatalysts (enzymes) are mainly applied for the production of fine chemicals when high (chemo-, regio- and stereo-) selectivity is required whereas inorganic catalysts (metals, metal oxides) are employed for large scales processes. In the latter case, high selectivity is desirable to save energy and natural resources [1].

The understanding of the reaction mechanism during a heterogeneously catalyzed reaction is still limited and recently, there is only one process (ammonia synthesis over iron catalysts) which can be claimed to be almost fully understood. However, the direct observation of the single elementary steps is not possible to date and the mechanism was derived from surface studies with iron single crystals [2]. Since the catalyzed reaction requires adsorption of the educts and intermediates and takes place on the surface of the catalyst, (*in-situ*) surface studies can be seen as one of the keys to a wider insight into the mode of operation of a catalyst. However, due to physical limitations, surface sensitive methods can often only be carried out under low pressures which are far away from *in-operando* conditions which usually require high pressures. The difficulty of transferring the obtained results is denoted as *pressure gap*. Thus, complementary methods of material science are needed to correlate atomic structure and macroscopic properties. With the acquired knowledge about these so-called structure function relationships, the conventional way of trial and error catalyst development can be shifted to a more rational catalyst design.

1.2 Methanol synthesis over Cu/ZnO/Al$_2$O$_3$ catalysts

1.2.1 Methanol

Methanol is one of the most important basic components in the chemical industry. The worldwide production volume was about 45 million tons in 2010, with a rising tendency over the past years [3]. The major amount is used to synthesize formaldehyde (precursor for organic synthesis), further products are methyl-tert-butyl-ether (antiknock agent) and acetic acid (precursor for monomers, conservation) (Figure 1-1).

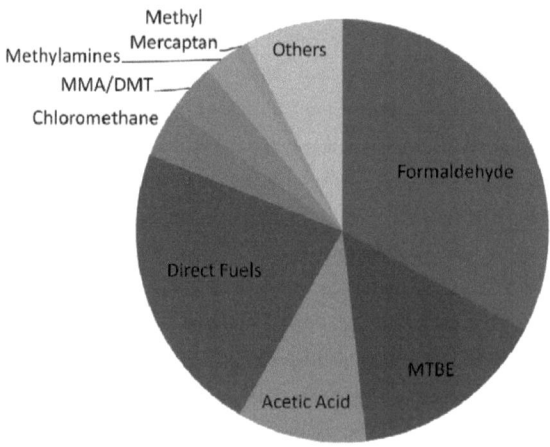

Figure 1-1: Distribution of global methanol consumption in 2010 [3]

Furthermore, a steadily increasing fraction and absolute amount of methanol is applied as alternative fuel acting as additive or for the production of biodiesel and dimethylether. The reasons are that methanol has a high energy content of 726.3 kJ mol^{-1} [4]. It shows good combustion properties and is a potential chemical H$_2$-carrier. One liter contains more hydrogen than pure hydrogen itself and only methane has an equally high H/C ratio. Advantages follow from the fact, that methanol is liquid at room temperature and therefore offers better transport and storing properties compared to elemental hydrogen. The freezing point is at around 176 K, so an application in cold regions can be performed unproblematically. On the other hand methanol requires resistant materials due to its polar character unlike nonpolar gasoline [5]. The combustion of methanol in engines proceeds almost without harmful byproducts like SO$_x$ or NO$_x$, such as being produced from impurities in gasoline. The application in fuel cells is not limited to the conventional hydrogen based type with previous steam reforming but also direct methanol fuel cells are possible [6]. Impurities like sulfur and produced CO from methanol steam reforming have to be removed to prevent poisoning the electrodes of the fuel cell [7].

In the chemical industry, methanol acts as a starting material to produce gasoline, olefins (for polymers) und aromates. Hence, it competes with the normally applied educts from coal, petroleum and natural gas, otherwise it promises a bigger independence from these fossil sources [8].

1.2.2 Methanol synthesis

Methanol synthesis [9] from synthesis gas (H_2, CO and CO_2) over solid catalysts was first reported in 1921 by Patart [10-11]. BASF launched the first large industrial methanol plant using ZnO/Cr_2O_3 catalysts, temperatures of 573-633 K and pressures of 150-250 bars [12]. In the 1960s, changing feedstock from coal to naphtha or natural gas led to less impurities (especially sulfur) in the synthesis gas and the known $Cu/ZnO/Al_2O_3$ catalysts were favored thenceforth applying somewhat lower temperatures and pressures up to 100 bars [13]. Nowadays, modern plants produce more than 5000 tons every day. Methanol synthesis from synthesis gas is exothermic ($\Delta H^0 < 0$) accompanied by decreasing entropy and therefore favored to proceed at low reaction temperatures and high reaction pressures. Thermodynamically, methanol is one of the least likely products compared to methane or higher alcohols. According to ref. [14], the following reactions are involved:

$$CO + 2\,H_2 \rightleftharpoons CH_3OH \qquad \Delta H^0 = -91 \text{ kJ mol}^{-1} \qquad \text{(Eq. 1.1)}$$

$$CO_2 + 3\,H_2 \rightleftharpoons CH_3OH + H_2O \qquad \Delta H^0 = -49 \text{ kJ mol}^{-1} \qquad \text{(Eq. 1.2)}$$

$$CO + H_2O \rightleftharpoons CO_2 + H_2 \qquad \Delta H^0 = -41 \text{ kJ mol}^{-1} \qquad \text{(Eq. 1.3)}$$

The hydrogenation-step can occur on both, CO (Eq. 1.1) or CO_2 (Eq. 1.2). The water-gas-shift reaction (Eq. 1.3) can be regarded as the difference of (Eq. 1.1) and (Eq. 1.2) and thus, is not an independent reaction. The mechanism and the carbon source of industrial methanol synthesis are still under debate today. In the early 1950's, a mechanism for methanol formation from CO over ZnO/Cr_2O_3 catalysts was reported [15]. Later, CO was assumed to be the predominant carbon source over Cu/ZnO catalysts [16]. CO_2 was only regarded to reoxidize Cu^0 to the active Cu^+ state. But at the same time, a Russian group reported methanol to be formed almost exclusively from CO_2. These results were obtained over $Cu/ZnO/Al_2O_3$ and were based on kinetic experiments [17-18] and radioactively labeled carbon dioxide isotopes [19-21]. No methanol was formed when using a pure CO/H_2 mixture over Cu/ZnO [22] or $Cu/ZnO/Al_2O_3$ [17]. Today, mechanism starting from CO_2 is mostly accepted when regarding industrial conditions.

Other synthesis routes for methanol formation start with CH_4 or pure CO_2. Both substances are known as "greenhouse gases", hence their economic conversion is desired but not possible to date [23]. Supported copper nanoparticles were not only widely applied as active catalysts in methanol synthesis but, depending on feed composition, temperature and pressure, also in methanol steam reforming and water-gas shift reaction because of the similar elemental reactions.

With a fundamental understanding of the reaction mechanism and correlations between catalytic activity and surface / bulk structure improved catalysts can be designed [24]. However, only for Cu single crystals the elemental steps of the methanol synthesis are known [25]. The questions concerning the microkinetics and the active centers remain still unanswered in the case of the more complex $Cu/ZnO/Al_2O_3$ catalytic systems. Reasons are the varying structure of Cu/ZnO catalysts accompanied by the change of the catalysts surface depending on the ambient conditions (oxidizing / reducing atmosphere) [26-28].

1.2.3 Preparation and Characteristics of Cu,Zn based catalysts

Preparation of $Cu/ZnO/Al_2O_3$ catalytic systems has been optimized in the last 40 years of industrial application. Different methods like co-precipitation, kneading, impregnation and leaching have been tested. Currently, most syntheses are carried out as a multi-step synthesis as follows [29]: Mixed metal hydroxy carbonate precursors are formed by controlled co-precipitation (pH 6.5, T = 338 K) from aqueous Cu,Zn,Al (6:3:1) nitrate solutions and Na_2CO_3 solution as precipitating agent. Chlorides and sulfates cannot be used because chloride and sulfur poison the final catalyst. Subsequently, the precipitate is aged in the mother liquor, filtrated, washed, dried and calcined to give the metal oxides. To obtain the desired activity in methanol synthesis, CuO has to be reduced to metallic copper. Normally, this takes place directly in the synthesis gas feed by hydrogen or carbon monoxide.

In order to better understand the catalytic systems, relationships between preparation parameters, microstructure and activity of $Cu/ZnO/Al_2O_3$ are investigated [30]. All parameters of the catalyst preparation influence the bulk and surface structure and therewith the characteristics and activity of the resulting catalyst. This phenomenon is also called the "chemical memory" [31].

Depending on the nominal metal composition, many different mixed metal hydroxy carbonate precursor phases can emerge in the course of catalyst preparation. With Cu as the major fraction, these are notably $Cu_2(OH)_2CO_3$ (malachite) for pure Cu samples, $(Cu_{1-x}Zn_x)_2(OH)_2CO_3$ (zincian malachite) with $x < 0.3$, $(Cu_{1-y}Zn_y)_5(OH)_6(CO_3)_2$ (aurichalcite) with $y > 0.5$, and $(Cu,Zn)_6Al_2(OH)_{16}CO_3 \cdot 4H_2O$ (hydrotalcite-like phase), only when a significant amount of Al^{3+} is present. The last phase should also be formed with other trivalent ions instead of Al^{3+}, such as Ga^{3+} or Cr^{3+}. Controversial discussions are present in the literature about the relevant precursor phase. Increased Cu dispersion, intrinsic activity (activity per Cu surface area) and overall activity were reported to be a consequence of the predominant presence of the precursors aurichalcite [32] or zincian malachite [33] in the Cu,Zn system and rosasite [34] in Cu,Zn,Al systems.

In further XRD studies, a shift of the $20\bar{1}$ (and $21\bar{1}$) reflection of zincian malachite has been observed when varying the Cu:Zn ratio [35-36]. This is related to the decrease of the d-spacing which is the result of incorporation of Zn^{2+} into the malachite structure (Figure 1-2).

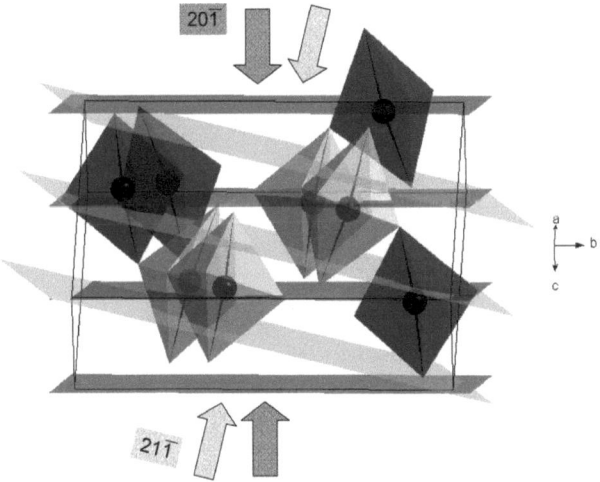

Figure 1-2: Unit cell of zincian malachite according to Behrens [37]. For clarification, only the Jahn–Teller elongated bonds of the CuO_6 units are shown. They are oriented either perpendicular to $(20\bar{1})$ (dark grey) or to $(21\bar{1})$ (light grey) and are contracted upon Cu/Zn substitution (see direction of arrows).

Cu^{2+} (3d^9) is a Jahn-Teller ion, Zn^{2+} (3d^{10}) is not. With increased Zn^{2+} incorporation, the elongated axial O-Cu-O units (perpendicular to the 20$\bar{1}$ netplanes) in the CuO$_6$ octahedra of pure Cu malachite are substituted by not elongated O-Zn-O units leading to a decrease of the average Jahn-Teller distortion. The shift is a direct measure for the incorporation of Zn^{2+} in the malachite structure and might predetermine the Cu dispersion and the activity of the resulting catalyst. The substitution of Cu^{2+} by Zn^{2+} in zincian malachite was found to be limited to approximately 28%. Higher Zn^{2+} contents would lead to the undesired Zn enriched phase aurichalcite [29, 31] (Figure 1-3).

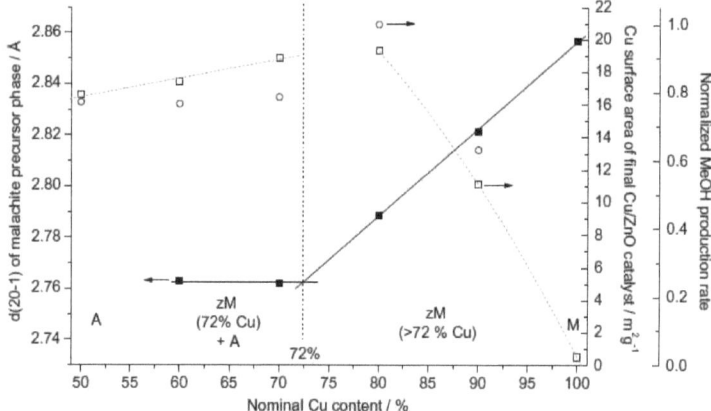

Figure 1-3: Graphic according to Behrens [33] showing d-spacing of the 20$\bar{1}$ netplanes of the zincian malachite precursor phase (■), Cu surface area (□) and normalized methanol production rates (○) of the final binary model catalyst as a function of nominal Cu content. The presence of the phases aurichalcite (A), zincian malachite (zM) and malachite (M) is indicated.

During precursor preparation, aging of binary Cu,Zn precipitates was reported to be crucial [29, 38-39] and to lead to a loss of the by-phase aurichalcite phase yielding more and higher Zn substituted zincian malachite (Figure 1-4). As a consequence, small and well distributed Cu and ZnO crystallites in the active catalyst lead to better performance in methanol synthesis. During aging the meta-stable product (amorphous precipitate) is transformed in the thermodynamic product (crystalline precipitate) by stirring in the mother liquor. The aging process has not been well understood yet and the proceeding reaction steps can hardly be optimized independently, because they are coupled to the synthesis parameters during co-precipitation.

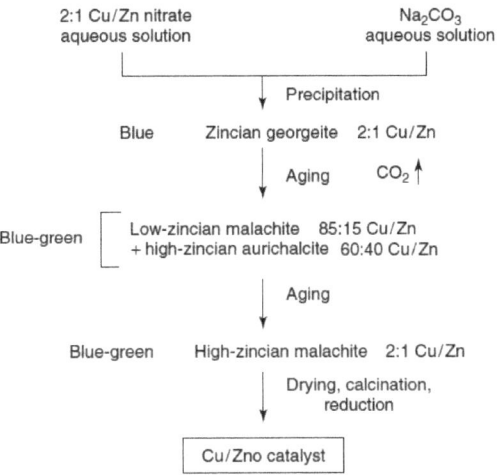

Figure 1-4: Proposed reaction scheme for precipitation, aging and subsequent stages in the preparation of 2:1 Cu/Zn catalysts. Graphic according to [9]; redrawn from [40].

A precondition for superior performance of Cu/ZnO/(Al$_2$O$_3$) catalysts in methanol synthesis is definitely a high Cu surface area due to the possibly increasing number of active sites. Whereas the activity of supported catalysts with low metal content is often referenced to the metal loading of the active phase, this does not work for the Cu/ZnO/(Al$_2$O$_3$) system because of the high Cu content where only a small fraction of the Cu atoms is accessible to the reaction gas. Linear correlations between Cu surface area and activity have been reported in literature for Cu,Zn [41] and Cu,Zn,Al system [30, 42]. However, deviations from this behavior have been observed depending on structural defects, i.e. microstrain [24, 43-46]. In general, a non-ideal form of copper [44, 47] is required to obtain active sites.

ZnO acts not only as dispersant and stabilizer but was also reported to cause synergetic effects and to provide the active centers due to the interface contact with the copper phase [39, 48]. That results in a beneficial electronic structure for adsorption of reactants and products. As a consequence, the activity of Cu/ZnO is several magnitudes higher than that of individual Cu or ZnO. Different active species have been proposed, e.g. Cu-Zn alloy formed during reduction [49], dissolved Cu$^+$ in ZnO [50] or electron rich Cu at Schottky-junctions [51]. In a recently published model of the active site of industrial methanol synthesis over Cu/ZnO/Al$_2$O$_3$, that was partially based on the work presented in this thesis, the synergetic effect was accounted for by strong metal support interaction (SMSI) [52], which has been observed in high-performance catalysts by HRTEM and XPS. Therein, the intrinsic activity of the exposed Cu surface area scaled with the abundance of stacking faults in Cu nanoparticles. This correlation was rationalized by the generation of high energy sites at the surface at the positions, where the planar defect terminates. Also residual oxygen in Cu as a result of incomplete reduction might play a role for the defect structure of active Cu. SMSI between Cu and ZnO has previously been reported in literature and studied by Cu surface area determination [42], EXAFS [27] and IR spectroscopy of CO adsorption [53].

For industrial applications of Cu/ZnO during methanol synthesis, mostly Al_2O_3 is used as a promoter. Al_2O_3 inhibits thermal sintering of the particles, prevents poisoning of the active metal surface and ensures additional chemical and thermal performance stability, which is very important for industrial catalysts [54-56]. Furthermore, addition of Al_2O_3 leads to higher intrinsic activities [42]. Other promoters like zirconia [57], silica [58], gallia and chromia [59] are also able to beneficially affect Cu dispersion, stability, activity and selectivity of the catalyst. Saito et al. [59] reported that addition of metal oxide promoters can have different effects, first the increase of the Cu dispersion in the case of alumina or zirconia, secondly the improvement of the specific activity in the case of gallia and chromia. The authors claim that the latter feature is due to the optimization of the Cu^+/Cu^0 ratio on the Cu surface under reaction conditions [60]. Recently, we found that an Al content of around 3 mol% in the ternary $Cu/ZnO/Al_2O_3$ system leads to an optimized beneficial promoting effect [61]. The obtained precursor during preparation was pure zinc (aluminum) containing malachite without any aurichalcite or Cu,Zn,Al hydrotalcite. After calcination, Al was introduced into the ZnO phase at Zn^{2+} sites in tetrahedral coordination. Alike ZnO, the function of the Al_2O_3 promoter was divided into a geometrical and a synergetic contribution. The former affects the Cu dispersion and leads to an increase of the Cu surface area. The latter promotes the intrinsic activity of Cu and was related to the incorporation of Al into the ZnO lattice and an influence onto the Cu/ZnO synergy.

The lifetime of industrial $Cu/ZnO/Al_2O_3$ catalysts lies in the range of years. Deactivation processes of Cu based catalysts mainly comprise sintering and poisoning. The first is diminished by the presence of ZnO and Al_2O_3 and can be regarded as a mechanical spacing effect [54, 62-64]. The latter can be alleviated by ZnO which absorbs sulfur (present as H_2S) from the feed gas [54, 64].

1.3 Outline of the work

This work contains systematic studies concerning the preparation and characterization of Cu,Zn-based catalytic systems for methanol synthesis. The binary Cu,Zn system presents a functional model system of the industrially applied Cu,Zn,Al system, but with less complexity. Correlations between preparation parameters, microstructure and catalytic activity in methanol synthesis shall be identified and investigated. Assuming zincian malachite as the relevant precursor phase [33], the preparation of phase pure samples with homogeneous metal distribution is targeted to enable unambiguous structure-function-relationships. Especially correlations of precursor chemistry (e.g. $d_{20\bar{1}}$ value of zincian malachite) with properties of the final catalyst (e. g. activity) are of great interest. Due to the "chemical memory", reproducibility of all preparation steps is an important factor.

One strategy is to perform systematic variation of preparation parameters for a system of fixed composition (here: Cu:Zn = 70:30). In order to better understand the aging process of the initial Cu,Zn precipitate during precursor preparation, this step is decoupled from co-precipitation and investigated independently with the help of *in-situ* energy dispersive X-ray diffraction. Application of different aging parameters like pH value and temperature shall reveal the influence on the process of precursor crystallization. Additionally, the pH value during co-precipitation is varied. Although studies for the ternary Cu,Zn,Al system are available [30, 65], data of comprehensive investigation of the less complex binary Cu,Zn system is still lacking.

The second strategy is to work with constant preparation conditions and apply modifications on the system. Promoting the binary Cu,Zn (70:30) system with small amounts of Al^{3+} was recently reported to lead to phase pure zincian malachite precursors with high Zn incorporation and subsequent higher activities [61]. The incorporation of Al^{3+} itself into the zincian malachite structure is possible but limited due to the charge mismatch. Similar results are expected when using Ga^{3+} as a promoter because of the not too different ionic radii of Al^{3+} and Ga^{3+} and analogous modifications of the oxides. The advantage of using Ga^{3+} instead of Al^{3+} is a better access to spectroscopic methods (XAS) which can account for elucidation of the promoter effect. Therefore, Cu,Zn,Ga samples with different Ga^{3+} concentrations up to 10 mol% are prepared.

Although the phenomenon of Cu-ZnO-synergy is controversially discussed in literature, the beneficial effect on the activity in industrial methanol synthesis is not questioned. We recently reported that the generation of "methanol copper" is induced by strong metal support interactions (SMSI) in the presence of ZnO [52]. If this image is right, then MgO cannot adequately replace ZnO, despite a possible enhancement of Cu dispersion. A combination of the effects of ZnO and MgO on Cu might lead to catalysts exhibiting a large and highly active Cu surface area.

Characterization data of the samples is presented from different stages of their preparation (precursors, calcined and reduced samples) with respect to crystalline phases (XRD), surface area (BET), thermal properties (TG-MS), reduction behavior (TPR), morphology and real composition (SEM, TEM, XRF) and Cu surface area (N_2O-RFC). Selected samples are subjected to further characterization (XAS) and testing in methanol synthesis. The process of aging is studied by in-situ energy dispersive XRD (University of Kiel, HASYLAB).

1.4 References

[1] J. A. Dumesic, G. W. Huber, M. Boudart, in *Handbook of Heterogeneous Catalysis, Vol. 6* (Eds.: G. Ertl, G. Knözinger, F. Schüth, J. Weitkamp), Wiley-VCH, Weinheim 2nd ed., **2008**, pp. 2501-2575.
[2] R. Schlögl, in *Handbook of Heterogeneous Catalysis* (Eds.: G. Ertl, G. Knözinger, F. Schüth, J. Weitkamp), Wiley-VCH, Weinheim 2nd ed., **2008**.
[3] www.methanolmsa.com, April **2012**.
[4] F. D. Rossini, *P. Natl. Acad. Sci. USA* **1931**, *17*, 343-347.
[5] G. A. Olah, A. Goeppert, G. K. Surya Prakash, *Beyond Oil and Gas: The Methanol Economy*, Wiley-VCH, Weinheim, **2006**.
[6] S. Ahmed, M. Krumpelt, *Int. J. Hydrogen Energ.* **2001**, *26*, 291-301.
[7] S. Velu, K. Suzuki, T. Osaki, *Catal. Lett.* **1999**, *62*, 159-167.
[8] U. Onken, A. Behr, *Chemische Prozesskunde, Lehrbuch der Technischen Chemie, Vol. 3*, Thieme, Stuttgart, **1996**.
[9] J. B. Hansen, P. E. H. Nielsen, in *Handbook of Heterogeneous Catalysis, Vol. 6* (Eds.: G. Ertl, G. Knözinger, F. Schüth, J. Weitkamp), Wiley-VCH, Weinheim 2nd ed., **2008**, pp. 2920-2949.
[10] G. Patart, in *French Patent 540*, **1921**, p. 343.
[11] C. Lormand, *Ind. Eng. Chem.* **1925**, *17*, 430-432.
[12] BASF, in *German Patents 415 686, 441 433, 462 837*, **1923**.
[13] ICI, in *GB Patents 1 010 871 (1961), 1 159 212 (1969), 1 296 212 (1972)*.
[14] M. R. Rahimpour, *Chem. Eng. Comm.* **2007**, *194*, 1638-1653.
[15] G. Natta, P. Pino, G. Mazzanti, I. Pasquon, *Chim. Ind. Milan* **1953**, *35*, 705-724.
[16] K. Klier, V. Chatikavanij, R. G. Herman, G. W. Simmons, *J. Catal.* **1982**, *74*, 343-360.
[17] A. Y. Rozovskii, Y. B. Kagan, G. I. Lin, E. V. Slivinskii, S. M. Loktev, L. G. Liberov, A. N. Bashkirov, *Kinet. Catal.* **1975**, *16*, 706-707.
[18] A. Y. Rozovskii, Y. B. Kagan, G. I. Lin, E. V. Slivinskii, S. M. Loktev, L. G. Liberov, A. N. Bashkirov, *Kinet. Catal.* **1976**, *17*, 1132-1138.
[19] Y. B. Kagan, G. I. Lin, A. Y. Rozovskii, S. M. Loktev, E. V. Slivinskii, A. N. Bashkirov, I. P. Naumov, I. K. Khludenev, S. A. Kudinov, Y. I. Golovkin, *Kinet. Catal.* **1976**, *17*, 380-384.

[20] A. Y. Rozovskii, G. I. Lin, L. G. Liberov, E. V. Slivinskii, S. M. Loktev, Y. B. Kagan, A. N. Bashkirov, *Kinet. Catal.* **1977**, *18*, 578-585.
[21] A. Y. Rozovskii, *Kinet. Catal.* **1980**, *21*, 78-87.
[22] B. Denise, R. P. A. Sneeden, *Appl. Catal.* **1986**, *28*, 235-239.
[23] K.-O. Hinrichsen, J. Strunk, *Nachr. Chem.* **2006**, *54*, 1080-1084.
[24] B. L. Kniep, T. Ressler, A. Rabis, F. Girgsdies, M. Baenitz, F. Steglich, R. Schlögl, *Angew. Chem. Int. Edit.* **2004**, *43*, 112-115.
[25] T. S. Askgaard, J. K. Norskov, C. V. Ovesen, P. Stoltze, *J. Catal.* **1995**, *156*, 229-242.
[26] N. Y. Topsoe, H. Topsoe, *Top. Catal.* **1999**, *8*, 267-270.
[27] J. D. Grunwaldt, A. M. Molenbroek, N. Y. Topsoe, H. Topsoe, B. S. Clausen, *J. Catal.* **2000**, *194*, 452-460.
[28] P. L. Hansen, J. B. Wagner, S. Helveg, J. R. Rostrup-Nielsen, B. S. Clausen, H. Topsoe, *Science* **2002**, *295*, 2053-2055.
[29] D. Waller, D. Stirling, F. S. Stone, M. S. Spencer, *Faraday Discuss.* **1989**, *87*, 107-120.
[30] C. Baltes, S. Vukojevic, F. Schüth, *J. Catal.* **2008**, *258*, 334-344.
[31] B. Bems, M. Schur, A. Dassenoy, H. Junkes, D. Herein, R. Schlögl, *Chem. Eur. J.* **2003**, *9*, 2039-2052.
[32] T. Fujitani, J. Nakamura, *Catal. Lett.* **1998**, *56*, 119-124.
[33] M. Behrens, *J. Catal.* **2009**, *267*, 24-29.
[34] R. H. Höppener, E. B. M. Doesburg, J. J. F. Scholten, *Appl. Catal.* **1986**, *25*, 109-119.
[35] M. Behrens, F. Girgsdies, A. Trunschke, R. Schlögl, *Eur. J. Inorg. Chem.* **2009**, 1347-1357.
[36] P. Porta, S. Derossi, G. Ferraris, M. Lojacono, G. Minelli, G. Moretti, *J. Catal.* **1988**, *109*, 367-377.
[37] M. Behrens, F. Girgsdies, *Z. Anorg. Allg. Chem.* **2010**, *636*, 919-927.
[38] M. S. Spencer, *Catal. Lett.* **2000**, *66*, 255-257.
[39] J. C. J. Bart, R. P. A. Sneeden, *Catal. Today* **1987**, *2*, 1-124.
[40] A. M. Pollard, M. S. Spencer, R. G. Thomas, P. A. Williams, J. Holt, J. R. Jennings, *Appl. Catal. A* **1992**, *85*, 1-11.
[41] G. C. Chinchen, K. C. Waugh, D. A. Whan, *Appl. Catal.* **1986**, *25*, 101-107.
[42] M. Kurtz, N. Bauer, C. Buscher, H. Wilmer, O. Hinrichsen, R. Becker, S. Rabe, K. Merz, M. Driess, M. A. Fischer, M. Muhler, *Catal. Lett.* **2004**, *92*, 49-52.
[43] E. N. Muhamad, R. Irmawati, Y. H. Tautiq-Yap, A. H. Abdullah, B. L. Kniep, F. Girgsdies, T. Ressler, *Catal. Today* **2008**, *131*, 118-124.
[44] P. Kurr, Technical University Berlin **2008**.
[45] M. M. Günter, T. Ressler, B. Bems, C. Buscher, T. Genger, O. Hinrichsen, M. Muhler, R. Schlögl, *Catal. Lett.* **2001**, *71*, 37-44.
[46] T. Ressler, B. L. Kniep, I. Kasatkin, R. Schlögl, *Angew. Chem. Int. Ed.* **2005**, *44*, 4704-4707.
[47] I. Kasatkin, P. Kurr, B. Kniep, A. Trunschke, R. Schlögl, *Angew. Chem. Int. Edit.* **2007**, *46*, 7324-7327.
[48] R. Burch, R. J. Chappell, S. E. Golunski, *Catal. Lett.* **1988**, *1*, 439-443.
[49] Y. Kanai, T. Watanabe, T. Fujitani, M. Saito, J. Nakamura, T. Uchijima, *Catal. Lett.* **1994**, *27*, 67-78.
[50] R. G. Herman, K. Klier, G. W. Simmons, B. P. Finn, J. B. Bulko, T. P. Kobylinski, *J. Catal.* **1979**, *56*, 407-429.
[51] J. C. Frost, *Nature* **1988**, *334*, 577-580.
[52] M. Behrens, F. Studt, I. Kasatkin, S. Kühl, M. Hävecker, F. Abild-Pedersen, S. Zander, F. Girgsdies, P. Kurr, B.-L. Kniep, M. Tovar, R. W. Fischer, J. K. Nørskov, R. Schlögl, *Science* **2012**, *336*, 893-897.
[53] R. N. d'Alnoncourt, X. Xia, J. Strunk, E. Löffler, O. Hinrichsen, M. Muhler, *Phys. Chem. Chem. Phys.* **2006**, *8*, 1525-1538.
[54] M. V. Twigg, M. S. Spencer, *Top. Catal.* **2003**, *22*, 191-203.

[55] M. Kurtz, H. Wilmer, T. Genger, O. Hinrichsen, M. Muhler, *Catal. Lett.* **2003**, *86*, 77-80.
[56] M. V. Twigg, M. S. Spencer, *Appl. Catal. A* **2001**, *212*, 161-174.
[57] C. Yang, Z. Y. Ma, N. Zhao, W. Wei, T. D. Hu, Y. H. Sun, *Catal. Today* **2006**, *115*, 222-227.
[58] E. K. Poels, D. S. Brands, *Appl. Catal. A* **2000**, *191*, 83-96.
[59] M. Saito, T. Fujitani, M. Takeuchi, T. Watanabe, *Appl. Catal. A* **1996**, *138*, 311-318.
[60] T. Fujitani, M. Saito, Y. Kanai, T. Kakumoto, T. Watanabe, J. Nakamura, T. Uchijima, *Catal. Lett.* **1994**, *25*, 271-276.
[61] M. Behrens, S. Zander, P. Kurr, N. Jacobsen, J. Senker, G. Koch, T. Ressler, R. W. Fischer, R. Schlögl, in *Performance Improvement of Nano-Catalysts by Promoter-Induced Defects in the Support Material: Methanol Synthesis over Cu/ZnO:Al*, submitted to *J. Am. Chem. Soc.*
[62] J. Sloczynski, *Chem. Eng. Sci.* **1994**, *49*, 115-121.
[63] I. Løvik, PhD Thesis, Norwegian University of Science and Technology **2001**.
[64] M. S. Spencer, M. V. Twigg, in *Ann. Rev. Mater. Res., Vol. 35*, **2005**, pp. 427-464.
[65] J. L. Li, T. Inui, *Appl. Catal. A* **1996**, *137*, 105-117.

Chapter 2: In-situ EDXRD Study of the Chemistry of Aging of Co-precipitated Mixed Cu,Zn Hydroxycarbonates – Consequences for the Preparation of Cu/ZnO Catalysts

Stefan Zander, Beatrix Seidlhofer, Malte Behrens

Abstract

In order to better understand the critical influence of the synthesis parameters during preparation of Cu/ZnO catalysts at the early stages of preparation, the aging process of mixed Cu,Zn hydroxide carbonate precursors was decoupled from the precipitation and studied independently under different conditions, i.e. variations in pH, temperature and additives, using *in-situ* energy-dispersive XRD and *in-situ* UV-Vis spectroscopy. Crystalline zincian malachite, the relevant precursor phase for industrial catalysts, was formed from the amorphous starting material in all experiments under controlled conditions by aging in solutions of similar composition to the mother liquor. The efficient incorporation of Zn into zincian malachite can be seen as the key of Cu/ZnO catalyst synthesis. Two pathways were observed: Direct co-condensation of Cu^{2+} and Zn^{2+} into Zn-rich malachite at $5 \geq pH \geq 6.5$, or simultaneous initial crystallization of Cu-rich malachite and a transient Zn-storage phase. This intermediate re-dissolved and allowed for enrichment of Zn into malachite at $pH \geq 7$ at later stages of solid formation. The former mechanism generally yielded a higher Zn-incorporation. On the basis of these results, the effect of synthesis parameters like temperature and acidity are discussed and their effect on the final Cu/ZnO catalyst can be rationalized.

Chapter 2: In-situ EDXRD Study of the Chemistry of Aging of Co-precipitated Mixed Cu,Zn Hydroxycarbonates – Consequences for the Preparation of Cu/ZnO Catalysts

2.1 Introduction

Due to the enormous economical relevance of solid catalysts in chemical industry,[1] their skillful synthesis and phenomenological optimization often is far more advanced than the understanding of the rationale behind the resulting individual values of synthesis parameters. Modern analytical methods can help to develop phenomenological catalyst synthesis towards knowledge-based design. The Cu-based methanol synthesis catalyst is a prominent example for this evolution.

Binary Cu/ZnO samples (Cu:Zn ca. 70:30) serve as a model system for the industrially applied Cu/ZnO/Al_2O_3 catalyst, which contains ca. 5-10 mol% Al_2O_3 as a structural promoter. Performance of the catalysts scales linearly with the accessible Cu surface area, but only within certain families of Cu/ZnO catalyst, which were prepared by a similar method, e.g. by co-precipitation or from citric acid melts.[2] This observation highlights the crucial influence of the synthesis route on the catalytic properties of Cu/ZnO,[3] which is also termed the "chemical memory" of the system.[4] The difference among the material families are attributed to intrinsic promoting effects. Cu dispersion and intrinsic activity are beneficially influenced by the presence and homogeneous distribution of ZnO in the catalyst. Firstly, it stabilizes small Cu nanoparticles acting as a geometrical spacer between them.[5] Secondly, strong metal-oxide interactions between ZnO and Cu are assumed to contribute to the *in-situ* formation of catalytically active sites. Different models for this latter synergetic effect are discussed in literature.[6-13]

The most successful and industrially applied synthesis route of Cu/ZnO catalysts follows a multi-step procedure in which mixed metal hydroxide carbonate precursors are formed by controlled co-precipitation from aqueous Cu/Zn/(Al) nitrate solutions using soda solution as precipitating agent.[14] Subsequently, the precipitate is aged in the mother liquor, filtrated, washed, dried, calcined and finally reduced to yield the active catalyst. It is described in literature, that aging is a crucial step during synthesis of the precursor and that it is essential for preparation of a successful catalyst.[4, 14-17]

In the following, we will discuss the influence of aging conditions on the properties of the catalyst on the basis of the recently published model of hierarchical meso- and nano-structuring of industrial methanol synthesis catalysts, which explains the benefit of the hydroxide carbonate precursor method for preparation of Cu/ZnO catalyst (Figure 2-1a).[18] In brief, the co-precipitate undergoes two micro-structure directing steps during preparation. Firstly, a mixture of zincian malachite crystallizes from the initially amorphous co-precipitate zincian georgeite[19] during aging, both with the elemental formula $(Cu,Zn)_2(OH)_2(CO_3)$.[18] This step is associated with a minimum in pH and a color change from blue to bluish green (Figure 2-1b). Small amounts of aurichalcite, $(Cu,Zn)_5(CO_3)_2(OH)_6$ are often observed as a side-phase. Further aging was reported to lower the fraction of the aurichalcite phase in favor of zinc enriched malachite.[14, 16] Crystallization of zincian malachite occurs preferably in form of very thin and interwoven needles, which leads to the proper porous meso-structure.

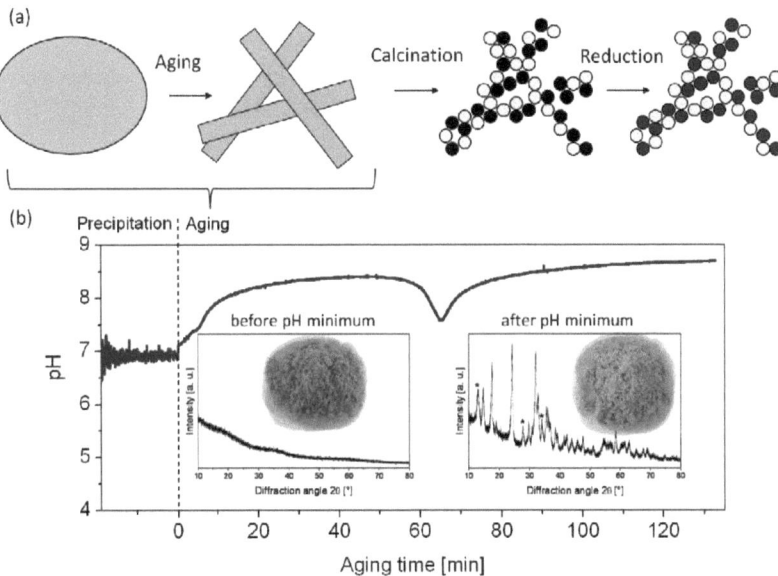

Figure 2-1: (a) Cartoon of the preparation of Cu/ZnO catalysts comprising precipitation of zincian georgeite, aging to form zincian malachite (meso-structuring), decomposition in CuO/ZnO aggregates (nano-structuring) and activation by reduction to Cu/ZnO. (b) pH evolution during precipitation and aging of a typical binary sample with change in sample color and crystallinity during aging (insets) The marked reflections in the XRD pattern refer to the aurichalcite by-phase, $(Cu,Zn)_5(CO_3)_2(OH)_6$. All other reflections are due to zincian malachite $(Cu,Zn)_2(CO_3)(OH)_2$.

Secondly, the nano-structuring of the individual precursor needles upon thermal decomposition yields aggregates of CuO and ZnO nanoparticles. Thus, the hierarchical pore structure of the final catalyst is already predetermined at the stage of the precursor. Here, the Zn concentration in zincian malachite needles is the crucial parameter, since significant amounts of atomically distributed Zn in the joint cationic lattice of zincian malachite lead to an effective stabilization of the Cu phase in high dispersion in the decomposition product (Figure 2-1a).[18] Due to solid state chemical constraints, the minimal Cu:Zn ratio in the zincian malachite phase is near 70:30.[20] For an efficient nano-structuring, the highest possible fraction of the available Zn amount should be incorporated in the zincian malachite precursor during aging.

The Zn fraction in this phase can be determined from the peak position of the $20\bar{1}$ reflection in the XRD pattern of zincian malachite. A low corresponding d-spacing is indicative of a high Zn content, which can be explained by a gradual contraction of this net plane distance caused by the average lowering of Jahn-Teller distortions of the octahedral MO_6 building blocks in malachite as Cu^{2+} is gradually replaced by Zn^{2+}.[18, 21] The shift of the $20\bar{1}$ reflection in the XRD pattern is, thus, a direct measure of the desired incorporation of Zn^{2+} into the malachite structure and serves as an estimate of the Cu dispersion in the final catalyst.

Hence, aging, i.e. the period of crystalline phase formation of the precursor, plays a key role for catalyst preparation and for the so-called chemical memory of Cu/ZnO catalysts. However, the effects of synthesis parameters like pH, temperature or mother liquor composition on the precipitate are not well understood and are so far related to the catalytic performance of the resulting Cu/ZnO catalyst only in a merely phenomenological manner. This lack of understanding can be seen as a major hindrance for further rational optimization of the Cu/ZnO/(Al$_2$O$_3$) system and requires a systematic and fundamental study of the chemistry of precipitate aging. Such a study is complicated by the fact that variation of a given parameter affects upstream precipitation as well as aging. The ambiguity if an observed change in the properties of the precipitate is a result of modified chemistry of aging or of changes in the precipitation process (resulting in a different starting material for downstream aging), requires experimental decoupling of both events. Furthermore, application of *in-situ* methods is desirable to ensure complete monitoring of all transformations happening during aging of the co-precipitate. *In-situ* energy-dispersive X-ray diffraction (EDXRD) has been shown to be a powerful method to study the mechanism[22] and kinetics[23] of solid state reactions,[24-25] e.g. under hydrothermal conditions[26] or in intercalation/de-intercalation reactions.[27]

In this paper we report a novel approach for the investigation of the aging process during preparation of Cu/ZnO catalysts using decoupled precipitation and aging steps and *in-situ* EDXRD and UV-Vis spectroscopy to monitor the latter.

2.2 Experimental

2.2.1 Precursor Preparation

Decoupling of precipitation and aging was realized by continuously feeding the initial amorphous co-precipitate slurry directly into a spray-dryer in order to suppress aging by fast drying. This "quenching technique" was necessary due to the fact that zincian georgeite is quite unstable in the mother liquor against crystallization. In course of the preparation, constant pH co-precipitation was performed in an automated laboratory reactor (Mettler-Toledo LabMax, 2 L, prefilled with 400 mL water) at T = 338 K and pH 7 from aqueous 1.6 M Na_2CO_3 solution and 1 M aqueous metal nitrate solution (Cu:Zn = 70:30). It is noted that the conditions of co-precipitation correspond to the conventional preparation process described in literature and were similar to the conditions of industrial catalyst preparation. A graphical representation of the precipitation log file can be found as supporting information (Figure S2-1). The resulting slurry was continuously removed from the co-precipitation reactor at the rate of addition of solutions (23 mL/min) and directly spray-dried (Niro Minor Mobile, T_{inlet} = 473 K, T_{outlet} = 373 K) after an estimated residence time of less than 20 min in the reactor and the connecting tubes. This is well below the aging period necessary for crystallization of the precursor material considering that pH minimum and color change are not expected before ca. 30 minutes of stirring in the mother liquor under these conditions.[18] Thus, the dried, solid product was X-ray amorphous except for some $NaNO_3$ resulting from crystallization of the counter ions during spray-drying (not shown). To remove $NaNO_3$ the precursor was thoroughly washed with cold water and spray-dried again leading to completely X-ray amorphous zincian georgeite. The Cu:Zn ratio of the solid was confirmed to be 73:27 (± 2%) by XRF. The resulting precursor is referred to as "unaged" despite its residence time of 20 min in the mother liquor because of the fact that it still was amorphous. Using this procedure, which is schematically summarized in Figure S2-2, we were able to employ a batch of unaged zincian georgeite as identical starting material for aging experiments under different conditions (T, pH, additives) in mother liquor-analogous media without affecting the co-precipitation process.

2.2.2 In-situ EDXRD and UV-Vis Spectroscopy during Simulated Aging

In order to simulate the aging process, 200 mg of the precursor were suspended in 2 mL of aging solution in a glass tube (internal diameter: 10 mm; volume: 7 mL). To keep the concentrations of the relevant ions in the aging solution near to the concentrations in the real mother liquor, it was freshly prepared by mixing appropriate amounts of the basic precipitating agent (1.6 M alkaline carbonate solution A_2CO_3, A = Na, K) and HNO_3 of a concentration corresponding to that of the mixed metal nitrate solution (1 M) until the desired pH was reached and stable. The *in-situ* measurements were started directly after preparation of the suspension. The uncovered glass tube was placed into a metal block, whose temperature was controlled by an oil bath. The suspension inside the glass tube was stirred during the aging experiment using a magnetic stir bar.

All *in-situ* aging investigations were carried out at the beamline F3 at HASYLAB/DESY, Hamburg, Germany. The beamline station receives white synchrotron radiation from a bending magnet with a critical energy of 16 keV and gives a positron beam energy of 4.5 GeV allowing detection of an energy range from 10 to 60 keV with a maximum in intensity at about 20 keV. An energy dispersive germanium detector was used to monitor the diffracted beam after transmission through the sample at a fixed angle, which was chosen as approximately 3.6° covering a d-spacing range of 2.6 to 12.2 Å. The beam was collimated to 100 × 100 μm. An acquisition time of 120 s yielded time-resolved X-ray powder patterns with sufficient counting statistics. The time span from placing the sample in the sample holder and start of recording the first diffraction patterns was less than 60 sec. The resulting spectra were evaluated using the EDXPowd[28] program package. More details on the experimental setup used can be found in the supporting information (Figure S2-3) and literature.[29] Phase evolution was followed by plotting the integral intensity of selected well-resolved reflection as a function of time. Additionally, the change of the color of the samples was tracked by UV-Vis spectroscopy in order to monitor the conversion of blue amorphous zincian georgeite to green crystalline zincian malachite. Diffuse reflectance measurements were performed with an OceanOptics optical fiber probe placed in the suspension well above the synchrotron beam. The probe was connected with a TopSensorSystems halogen lamp and an OceanOptics high resolution spectrometer HR2000CG-UV-NIR. The acquisition time was set to 120 s per spectrum

2.2.3 Ex-situ Characterization

All samples subjected to EDXRD measurements were cooled to room temperature after the *in-situ* experiments within 5 minutes, filtrated and washed with water. Conventional X-ray diffraction (XRD) measurements were performed with a STOE STADI P transmission diffractometer equipped with a primary focusing Ge monochromator (Cu-K$_{α1}$ radiation) and position-sensitive detector to determine the peak positions more accurately than was possible with EDXRD. All XRD patterns are presented as supporting information (Figure S2-4). The samples were mounted in the form of a clamped sandwich of small amounts of powder fixed with a small amount of grease between two layers of thin polyacetate film. Refinements were done in the 2θ range 4-80° using the software package TOPAS.[30] Domain sizes were determined from the XRD peak widths and are given as volume weighted mean column heights. Surface area determination was performed in a Quantachrome Autosorb-6 machine by N_2-adsorption-desorption using the BET method. Cu:Zn ratios of the samples were obtained from X-ray fluorescence (XRF) measurements using a Bruker S4 Pioneer X-ray spectrometer.

2.3 Results and Discussion

2.3.1 General

Using the method of sample preparation described in the experimental section allowed simulating the aging process of an amorphous binary zincian georgeite co-precipitate under controlled conditions similar to those used in course of preparation of industrial methanol synthesis catalysts (Figure 2-1). *In-situ* EDXRD and UV-Vis measurements allowed insight into the chemistry of aging, which is of crucial importance for the phase formation of the catalyst precursor and, thus, for the preparation of highly active catalysts.

The catalyst precursor was aged at different temperatures (323-343 K), starting acidities (pH 5.0–8.0) and using different counter cations (Na^+ and K^+). Crystalline zincian malachite $(Cu,Zn)_2(OH)_2(CO_3)$ was finally detected by XRD after aging for all conditions applied (supporting information, Figure S2-4). It is interesting to note that minor amounts of aurichalcite are typically observed for the Cu:Zn ratio of 70:30, e.g. after aging in a conventional 2-L batch reactor at 338 K and pH of 7.0 (see also marked reflections in Figure 2-1b).[18] The presence of aurichalcite was barely detectable in the samples after simulated aging by *ex-situ* XRD (supporting information, Figure S2-4). Inclusion of the aurichalcite phase in the Rietveld fits led to improved R-values for some samples and resulted in varying amounts of aurichalcite between 0 and 13 wt.% (Table 2-1). However, due to the poor crystallinity of the sample and the low amount of this phase the error is estimated to be at least ± 5 wt.%. A typical graphical representation of a typical Rietveld fit is given in Figure 2-2. In addition to the small differences in phase composition among the aged samples, variations in crystallinity, Zn content of the zincian malachite phase and specific surface area as a result of different aging conditions are reflected in the XRD domain sizes scattering between 9.2 and 10.9 nm, the $d_{20\bar{1}}$ value ranging between 2.757 and 2.775 Å and the BET surface areas being between 69 and 85 m^2g^{-1} (Table 2-1). These observations confirm the sensitivity of relevant properties of the Cu,Zn precursor to the exact conditions of crystallization for the same starting material and a systematic discussion will be given in the following.

2.3.2 Phase Evolution

The phase evolution during precursor aging will be discussed for the experiment conducted at T = 323 K and pH 7.0 using a Na^+ containing aging solution (ID 7 in Table 2-1). Despite the X-ray amorphous dry starting material, some very weak XRD peaks are already observed in the first *in-situ* pattern of the slurry recorded after less than 180 sec after starting the experiment (Figure 2-3a). After an aging time of some minutes, a steep increase in Bragg peak intensity is observed and two phases can be clearly distinguished (Figure 2-3b): Sodium zinc carbonate, $Na_2Zn_3(CO_3)_4 \cdot 3H_2O$,[31] and the target material zincian malachite, $(Cu,Zn)_2(OH)_2(CO_3)$.[32] At the end of the experiment, only zincian malachite was observed as the final product (Figure 2-3c). Sodium zinc carbonate $Na_2Zn_3(CO_3)_4 \cdot 3H_2O$ has been reported before in literature in the context of Cu/ZnO catalyst preparation. It was identified as the initial precipitate in Zn-rich or pure Zn systems, which upon aging transformed into aurichalcite or hydrozincite.[15, 33] In one study, it has probably also been detected in a Cu-rich system as a transient phase during aging,[34] but was assigned as "crystalline zincian georgeite".

Table 2-1: Summary of aging parameters and aging results. The sample ID 0 refers to the unaged precursor.

	Aging Conditions			*In-situ* results (EDXRD data)		*Ex-situ* results (recovered samples after EDXRD measurement)				
ID	pH	T [K]	A^+ in A_2CO_3	Na,Zn Intermediate	Onset [min]	Reaction Time [min][a]	Aurichalcite [wt%] ($\pm$ 5 wt%)	$d_{20\bar{1}}$ [Å]	Zn in $zM^{[b]}$ [%]	BET [m²/g]	FWHM- $LVol^{[c]}$ [nm]	Cu:Zn XRF [mol%] ($\pm$ 2 mol%)
0	-	-	-	-	-	-	-	-	-	15	-	73.3:26.7
1	5	333	Na^+	-	20	-	6	2.757	29.2	85	9.9	73.2:26.8
2	6	333	Na^+	-	34	-	8	2.759	28.5	81	9.7	72.3:27.7
3	6.5	333	Na^+	-	36	-	8	2.760	28.4	83	10.0	72.8:27.2
4	7	333	Na^+	x	12	30	0	2.767	26.2	82	9.6	73.8:26.2
5	7.5	333	Na^+	x	12	34	0	2.767	26.3	80	9.8	72.1:27.9
6	8	333	Na^+	x	14	34	0	2.768	26.0	83	9.6	72.5:27.5
7	7	323	Na^+	x	24	98	0	2.775	23.8	81	9.2	73.9:26.1
8	7	343	Na^+	x	6	16	0	2.765	26.7	72	10.9	73.9:26.1
9	7	333	K^+	-	56	-	13	2.767	26.2	77	9.8	72.3:27.7
10	7	343	K^+	-	18	-	11	2.775	23.8	69	10.0	72.0:28.0

[a] Time interval between appearance and complete consumption of the intermediate.
[b] Zn content in zincian malachite (zM) calculated from $d_{20\bar{1}}$ values; see also Figure 2-7.
[c] Crystallite sizes of zincian malachite determined from the half width of the XRD peaks using the TOPAS refinement software.

One suitable, well resolved peak of both phases was chosen for further EDXRD data evaluation. The phase fraction of zincian malachite was represented by the integral intensity of the $20\bar{1}$ peak, sodium zinc carbonate by the 222 peak. The phase evolution with time is shown in Figure 2-4a. The sodium zinc carbonate phase crystallizes in parallel to the zincian malachite phase and re-dissolves upon prolonged aging. The onset of crystallization occurs after 24 min and the re-dissolution of the sodium zinc carbonate occurs over 98 min without a significant increase in the zincian malachite phase.

The amorphous starting material ("zincian georgeite") is hard to comprehensively characterize. IR-spectroscopic studies of the unaged material have shown the presence of both hydroxide as well as carbonate anions.[4, 18] Here, we describe the starting precipitate as an amorphous double-salt with unknown anionic composition: $a\text{-}Cu_{0.7}(OH)_x(CO_3)_{0.7-x/2} \cdot Zn_{0.3}(OH)_y(CO_3)_{0.3-y/2}$. In the presence of varying amount of H_2O, OH^-_{aq} and $CO_3^{2-}_{aq}$ we can write for the two steps of the aging reaction:

$$a\text{-}Cu_{0.7}(OH)_x(CO_3)_{0.7-x/2} \cdot Zn_{0.3}(OH)_y(CO_3)_{0.3-y/2} + z\ Na^+_{aq} \quad \text{(Eq. 2-1)}$$
$$\rightarrow (Cu_{>0.7}Zn_{<0.3})_2(OH)_2CO_3 + z/2\ Na_2Zn_3(CO_3)_4 \cdot 3H_2O$$
$$\rightarrow (Cu_{0.7}Zn_{0.3})_2(OH)_2CO_3 + z\ Na^+_{aq}$$

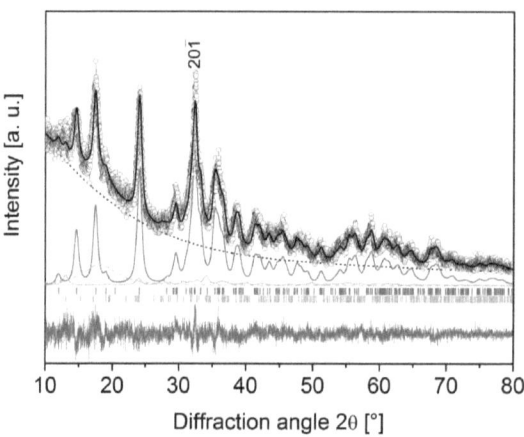

Figure 2-2: Rietveld refinement of the *ex-situ* XRD pattern of the sample aged *in-situ* at pH 6.5 and 333 K (ID 3) for quantitative analysis of the phase composition. Experimental data (circles), background (dotted), background peak (dashed, due to the grease used as sticking agent to keep the sample in place on the sample holder), calculated pattern zincian malachite (grey), calculated pattern aurichalcite (light grey), total calculated curve (black) and difference curve (grey, offset -100). This plot is representative for the other aged samples. In this case the ratio of zincian malachite to aurichalcite was calculated to be 92% to 8%.

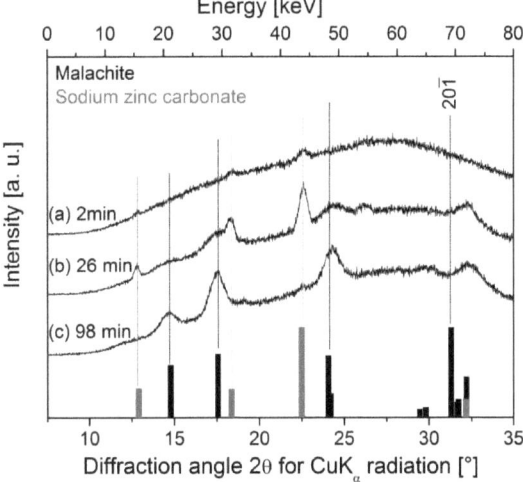

Figure 2-3: EDXRD patterns (converted to °2θ values of Cu Kα radiation) during aging of the amorphous precursor at pH 7 and 323 K (ID 7) after two (a), 26 (b) and 98 min (c). At the bottom PDF 72-75 (black bars) and PDF 1-457 (grey bars) are shown as references for zincian malachite $(Cu,Zn)_2(CO_3)(OH)_2$ and for sodium zinc carbonate $Na_2Zn_3(CO_3)_4 \cdot 3\ H_2O$, respectively. The *in-situ* EDXRD spectra are representative for all conducted experiments. The position of the $20\bar{1}$ peak of zincian malachite is shifted compared to the pure malachite reference because of zinc incorporation (see text).

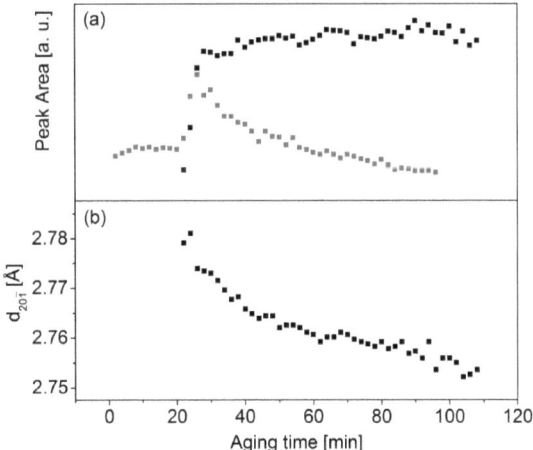

Figure 2-4: Integral intensity of selected EDXRD peaks vs. aging time (a); in Na_2CO_3; pH = 7; T = 323 K (ID 7). Zincian malachite, $(Cu,Zn)_2(OH)_2(CO_3)$, is represented by the $20\bar{1}$ peak (black), sodium zinc carbonate $Na_2Zn_3(CO_3)_4 \cdot 3\ H_2O$ by the 222 peak (grey). Corresponding d-spacing of the $20\bar{1}$ peak of zincian malachite (b).

If we assume $Na_2Zn_3(CO_3)_4 \cdot 3H_2O$ to be a pure Zn-phase with no Cu incorporation, the initially formed zincian malachite should be poor in Zn. In the following aging step, the sodium zinc salt acts as a Zn storage phase slowly deliberating its Zn content by dissolution. This fraction of Zn can either re-precipitate in form of a Zn-phase not detectable by XRD, or – as proposed in equation (1) – it can be incorporated into zincian malachite by re-crystallization increasing the Zn-content of this phase. The latter possibility is supported by the evolution of the $20\bar{1}$ peak position (Figure 2-4b). As the Zn storage phase re-dissolves, the peak is shifted to a lower d-spacing, indicating further incorporation of Zn into zincian malachite. It is noted, however, that a final proof of this mechanism is still lacking as there is a peak overlap of the $20\bar{1}$ of zincian malachite around 32.8 °2θ (for Cu K$_\alpha$ radiation) and the 422 of the sodium zinc salt located at 32.2 °2θ according to PDF 1-457. The intensity ratio of these two peaks is around 5:1 in this stadium. Diminishing of the latter peak due to dissolution may alone result in an artificial profile shift to higher angles in the EDXRD patterns. Unfortunately, the quality of the *in-situ* EDXRD patterns is not sufficient for a whole pattern refinement. After all, the above made assumption of an intrinsic peak shift of the $20\bar{1}$ of zincian malachite seems reasonable, because no narrowing of the peak profile with time was observed, which would be associated with decrease of a shoulder (see supporting information, Figure S2-5). Furthermore, the intensity of the 422 of sodium zinc carbonate is only 20% of the most intensive reflection of that phase, while the $20\bar{1}$ of zincian malachite is the strongest reflection of this phase. In Figure 2-3a, where only the sodium zinc carbonate phase is present, no significant intensity due to the 422 can be seen at a position corresponding to ca. 32 °2θ for Cu K$_\alpha$, suggesting that the contribution of the overlapping to the peak position of the phase mixture has only a minor influence.

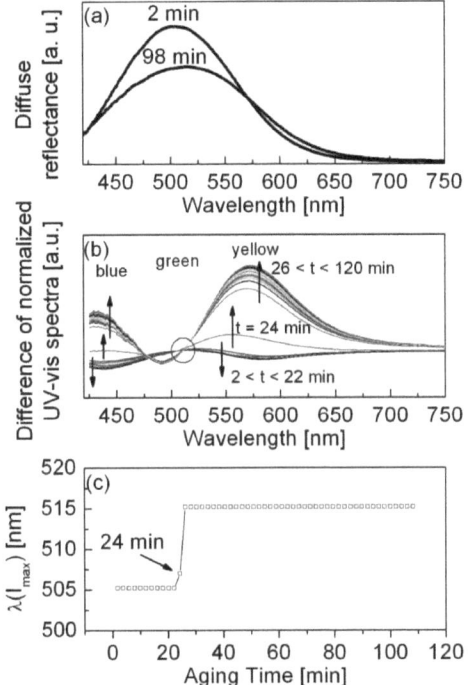

Figure 2-5: UV-Vis results for simulated aging in Na_2CO_3 at pH 7; 323 K (ID 7). a) Measured diffuse reflectance before and after aging. b) Difference of normalized spectra relative to the initial spectrum at t = 2 min. c) Wavelength of the maximum intensity in the UV-Vis spectra in the range of 425 to 900 nm as a function of time.

The UV-Vis diffuse reflectance spectra of the suspension corresponding to the starting material and the final product (aging conditions pH 7, 323 K, ID 7) are shown in Figure 2-5a. The change of the position of the broad signal from 505 to 515 nm reflects a change in crystal field splitting around the Cu^{2+} ions and the transition from blue to bluish green.[35] Difference plots of the normalized *in-situ* recorded spectra are shown in Figure 2-5b. It can be seen that several smaller bands contribute to the spectra. The presence of an isosbestic point near 510 nm was observed for all experiments and suggests that the starting material directly transforms into a single optically active product. This does not contradict the transient presence of the sodium zinc carbonate phase, but rather confirms the assumption that this phase does not contain Cu^{2+} ions and does not contribute to the reflectance in the Vis-range of the optical spectrum. The green part of the spectrum does hardly change and it can be seen that the change of color from blue to green is mostly due to an increase in reflectance in the yellow regime of the spectrum, which is much stronger compared to the increase in the blue part. The temporal evolution of the UV-Vis spectrum shows only minor changes in the beginning of the reaction up to aging times of 22 min. During this period a slight decrease of reflectance in the yellow and blue part is observed, which started directly as the starting material was in contact with the aging medium. In accordance with the EDXRD results, abrupt changes occur at an aging time of 24 min and reflectance in these parts of the spectrum sharply increases. Interestingly, the color change is finished almost immediately (after less than 4 min) and again only little changes are observed at 26 < t < 120 min, while the process of phase formation observed by EDXRD persists for 98 min. This clearly shows that the change of the color of the precursor slurry is not a suitable indicator for the end of the chemical changes happening during aging. UV-Vis spectroscopy probes changes on the molecular level, which naturally precede the detection with (ED)XRD technique as crystallization requires "oversaturation" of the newly formed complexes, which happens over a longer time scale under the conditions applied. In Figure 2-5c, the maxima of the broad reflectance signal are shown as a function of aging time, showing again the step-like change at the time of crystallization of malachite.

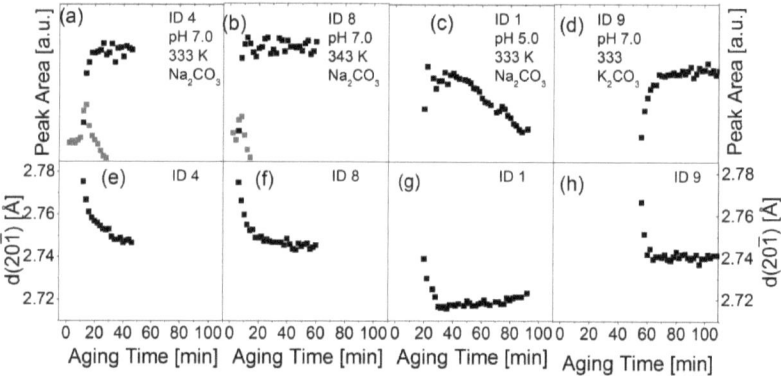

Figure 2-6: Measured features vs. aging time for samples noted in the plot (cf. Table 2-1). Top row a-d: Integral intensity of selected EDXRD peaks of zincian malachite $20\bar{1}$, (black) and sodium zinc carbonate 222, (grey). Bottom row e-h: Evolution of $d_{20\bar{1}}$ values of zincian malachite.

No residual sodium zinc carbonate or other by-phases were detected after 120 min of aging. In particular, no aurichalcite was detected in the *in-situ* EDXRD patterns. Aurichalcite might play a similar role as a Zn-uptake phase during aging.

Waller et al.[14] investigated the aging mechanism of a Cu:Zn = 67:33 system and observed that at first a mixture of crystalline zincian malachite (Cu:Zn ≈ 85:15) and aurichalcite (Cu:Zn ≈

60:40) were formed and subsequently transformed into zinc richer malachite (Cu:Zn ≈ 67:33) at the expanse of aurichalcite. We previously found that for conventional batch aging of a binary precursors (Cu:Zn = 70:30) for 2 hours at 338 K low amounts of the zinc richer phase aurichalcite co-exist with zincian malachite showing a Zn content of 27%.[18, 20] As mentioned above, low amounts of aurichalcite were detected by *ex-situ* XRD indicating that this phase indeed may act as a stable sink for Zn, but its amount probably is too little to be detected by *in-situ* XRD or that it has crystallized only upon drying of the samples.

2.3.3 The Effect of Temperature

At an aging temperature of 323 K, the sodium zinc carbonate storage phase was re-dissolved within 16 - 98 min upon aging at pH 7.0, depending on the temperature (Figure 2-4a and Figure 2-6a,b, Table 2-1, ID 7, 4, 8). The change of the color, assigned to the beginning formation of crystalline zincian malachite, always occurred within a few minutes and was tracked by UV-Vis-spectroscopy (Figure 2-5c).

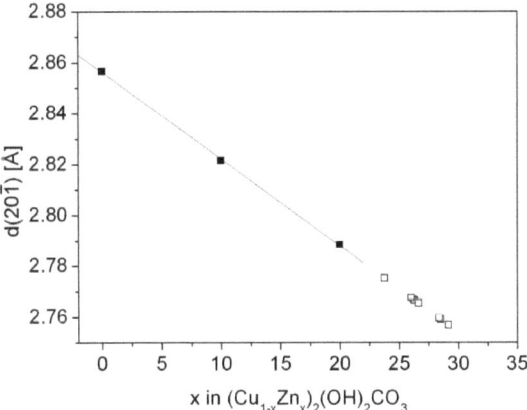

Figure 2-7: Calibration of the $d_{20\bar{1}}$ values versus the Zn content in zincian malachite. The three solid data points stem from reference samples described in ref. [18, 21] and were used for linear extrapolation. The open data points positioned onto the extrapolated line refer to the values observed in this study (from *ex-situ* XRD). The resulting Zn-contents are given in Table 2-1.

Increasing the temperature from 323 K to 333 or 343 K changed the kinetics of aging and led to earlier onset of crystallization and a shorter time period of existence of the sodium zinc phase (Figure 2-6a,b and Table 2-1, ID 4, 7 and 8), but no changes in the mechanism of aging were observed. The monotonous shift to lower $d_{20\bar{1}}$ values (Figure 2-6e,f) indicates the incorporation of Zn into zincian malachite with time after crystallization. *Ex-situ* XRD was applied on the recovered samples, which allows for a more accurate determination of the absolute peak positions than *in-situ* EDXRD due to higher instrumental resolution, longer acquisition time, more precise calibration and lower contribution of the background. The effect of temperature on the final degree of Zn incorporation in zincian malachite is reflected in the shift of the $d_{20\bar{1}}$ spacing (Table 2-1, ID 4, 7 and 8). While $d_{20\bar{1}}$ was similar for the higher temperatures, it was found to be significantly larger for the zincian malachite sample prepared at 323 K indicating a lower final degree of Zn incorporation at this temperature. This observation shows that, despite stemming from the same amorphous starting material, the degree of Zn incorporation can be affected by the aging conditions. The detrimental effect of low preparation and aging temperatures has been reported in the literature.[3, 36] A lack of Zn has been also observed for ternary $Cu/ZnO/Al_2O_3$ catalysts prepared at low pH or low temperature.[3] Our results suggest that this effect can be explained with a lack of Zn in the zincian malachite precursor phase. This is consistent with observations recently made during titration experiments[37] showing that the precipitation pH of Zn^{2+} is shifted to higher pH values as temperature decreases. Thus, the applied aging pH value may not be sufficiently basic to keep all Zn in the solid state at low temperatures and Zn^{2+} may be leached out of the precipitate at low temperatures and pH values. Interestingly, XRF measurements of the sample recovered after aging at different temperatures all showed the same average Cu:Zn ratio near 70:30 (Table 2-1). We thus conclude that the final pH is high enough to completely precipitate Zn^{2+} also at 323 K, but suggest that during the crystallization process, which is associated with an intermediate minimum in pH peaking roughly a full pH unit below the initial aging pH under these conditions[18] (Figure 2-1b), a transient leaching of Zn^{2+} from the starting material may occur at low temperatures during the pH minimum. This can explain a lack of Zn in the zincian malachite phase due to (partial) Zn dissolution at the time of its crystallization. Later re-precipitation leads to formation of low amounts of undetected Zn-rich phases resulting in the same average composition, but in an inhomogeneous and thus unfavorable Zn distribution in the product.

Thus, equation 1 has to be revised as $(Cu_{0.7}Zn_{0.3})_2(OH)_2CO_3$ is not an appropriate representation of the final product, which exhibits variations in its Cu:Zn ratio. We add an unknown Zn phase denoted Zn↓ as sink for "extra-lattice" Zn. The nature of this phase may be residual but undetected sodium zinc carbonate, amorphous or due to low abundance undetectable crystalline aurichalcite or another form of Zn-containing hydroxide or basic carbonate. As a function of the aging conditions this phase is present in various amounts and affects the Cu:Zn ratio in zincian malachite by limiting the available amount of Zn:

$$\text{a-Cu}_{0.7}(OH)_x(CO_3)_{0.7-x/2} \cdot \text{Zn}_{0.3}(OH)_y(CO_3)_{0.3-y/2} + z\, Na^+_{aq} \quad \text{(Eq. 2-2)}$$
$$\rightarrow (Cu_{\gg 0.7}Zn_{\ll 0.3})_2(OH)_2CO_3 + z/2\, Na_2Zn_3(CO_3)_4 \cdot 3H_2O$$
$$\rightarrow (Cu_{>0.7}Zn_{<0.3})_2(OH)_2CO_3 + z\, Na^+_{aq} + w\, Zn\downarrow$$

A quantification of the Zn-content in the final zincian malachite phase is possible by assuming a Vegard-type behavior of this lattice spacing and calibrating the obtained $d_{20\bar{1}}$ values with reference values from (zincian) malachite samples far from its limit of Zn incorporation. This is shown in Figure 2-7. If the data points are arranged on the extrapolated line according to their $d_{20\bar{1}}$ values, they cover variations in the Zn-content of zincian malachite between 23.8 and 29.2% Zn as a function of different aging conditions (Table 2-1). The largest Zn-contents determined with this method are very close to the nominal Cu:Zn-ratio applied during synthesis of 70:30, but exceed the Zn-content of the starting material determined by XRF. This discrepancy is attributed to the difference of the two methods and their calibration errors. In case of the sample obtained at 323 K the Zn content in zincian malachite is only 23.8%, while the rest of Zn is trapped in a relatively large amount of the Zn-sink phase Zn↓. At 333 or 343 K, the amount of Zn↓ is lower and the Zn-content in zincian malachite is 26.2 and 26.7%, respectively.

It is noted that a beneficial effect of lowering the temperature on the crystallite size is observed, which decreases with temperature from 10.9 nm at 343 K to 9.2 nm at 323 K. Accordingly, the lowest BET surface areas were found for the samples prepared at 343 K (Table 2-1).

In summary, the effect of increasing temperature accelerate the crystallization kinetics, leads to larger crystallites and, thus, is detrimental for the meso-structure of the catalyst precursor. Lowering the temperature to 323 K, however, leads to intermediate leaching of Zn^{2+} from the co-precipitate and to an unfavorable Zn distribution resulting in a lower degree of Cu,Zn-substitution of the zincian malachite phase. This hinders an effective nano-structuring of the catalyst. Thus, the empirically optimized aging temperature around 338 K can be understood from the chemistry of aging of the precursor and envisaged as the optimum of two antagonistic trends.

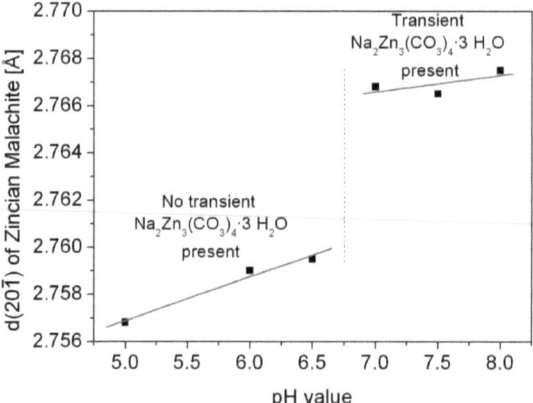

Figure 2-8: $d_{20\bar{1}}$ values of zincian malachite (from *ex-situ* XRD) depending on the pH value of simulated aging. Two different groups can be observed: The zincian malachite samples crystallized without intermediate formation of sodium zinc carbonate (pH 5-6.5) show low $d_{20\bar{1}}$ values indicating high Zn content. Crystallizations via the intermediate (pH 7-8) led to high $d_{20\bar{1}}$ values or low Zn content, respectively. Lines are guides for the eye.

2.3.4 The Effect of Acidity

The effect of different pH values during aging was investigated at 333 K with Na$^+$ containing aging solutions. The phase evolution data for aging at pH 7.0 is shown in Figure 2-6a, and only minor differences were observed if the experiment was conducted at a higher pH of 7.5 or 8.0 (Table 2-1, ID 4-6; supporting information: Figure S2-6). Also the crystallite sizes, specific surface areas and Zn contents of the resulting zincian malachite precursors were similar. Reducing the pH, however, strongly affected the mechanism of the aging process. Most striking is the absence of the sodium zinc carbonate phase at low pH associated with a delay of crystallization (Table 2-1, ID 1-3). Only for pH ≥ 7 sodium zinc carbonate was detected, while zincian malachite crystallized from the amorphous starting material at pH ≤ 6.5 without participation of any other EDXRD-detectable phase. The corresponding evolution of crystalline phases is presented in Figure 2-6c for aging at pH 5 and as supporting information for the other pH values (Figure S2-6). The decrease of intensity in Figure 2-6c is probably due to partial dissolution of zincian malachite at low pH value. Figure 2-6g shows that $d_{20\bar{1}}$ is constant or even slightly increasing directly after the crystallization period suggesting that there is hardly any change of the Zn-content in zincian malachite with aging time for these samples. Thus, a second, simpler mechanism of aging is present with only one detectable step:

$$\text{a-Cu}_{0.7}(\text{OH})_x(\text{CO}_3)_{0.7-x/2} \cdot \text{Zn}_{0.3}(\text{OH})_y(\text{CO}_3)_{0.3-y/2} \qquad \text{(Eq. 2-3)}$$

$\rightarrow (Cu_{>0.7}Zn_{<0.3})_2(OH)_2CO_3 + w\ Zn\downarrow$ (Eq. 3)

It can be seen from Table 2-1 that this mechanism of crystallization seems kinetically hindered as longer isothermal induction periods are required compared to reactions following the mechanism of equation 2 at the same temperature. The preferred Zn-sink phase $Zn\downarrow$ for this mechanism can be identified as aurichalcite (Table 2-1).

Interestingly, despite the presence of significant amounts of aurichalcite, there was a significantly higher degree of Zn-incorporation in the resulting zincian malachite if crystallized in absence of the sodium zinc carbonate according to equation 3. Accordingly, *ex-situ* XRD evaluation suggested the presence of two groups of precursors (Figure 2-8): The precursors crystallized without sodium zinc carbonate obtained at low pH with large amounts of Zn incorporated in the cationic lattice of zincian malachite (small $d_{20\bar{1}}$, 28.4-29.2% Zn in zincian malachite) and the ones obtained at higher pH with significantly lower amounts of Zn on Cu-sites (large $d_{20\bar{1}}$, 26.0-26.3% Zn in zincian malachite). The overall Cu:Zn ratio of the recovered solid detected by XRF was always near the starting composition for all samples (Table 2-1) suggesting again that other non-detectable Zn-rich by-phases $Zn\downarrow$ are present and act as a sink for Zn, in particular if crystallization occurred at high pH in presence of sodium zinc carbonate. The highest Zn incorporation of this study was detected for the samples obtained at T = 333 K and pH 5. According to Figure 2-7, the Zn content of zincian malachite is 29.2%. The successful minimization of any form of Zn segregation – like the transient crystallization of sodium zinc carbonate or the formation of the Zn-rich aurichalcite phase – helps to prepare a homogeneous precipitate capable of efficient nanostructuring during thermal decomposition according to the scheme presented in Figure 2-1a.

2.3.5 The Effect of Potassium Counter Ions

In case of the mechanism described by equation 2, it is tempting to relate the lack of Zn in the zincian malachite precursor to the amount of Zn, which has intermediately formed the sodium zinc salt. In order to test this hypothesis, analogous experiments were performed using an aging solution based on neutralized K_2CO_3 solution. The absence of Na^+_{aq} should suppress the formation of the transient sodium salt also at higher pH and therefore may have a promoting effect on the desired incorporation of Zn in zincian malachite at these conditions.

The experiments at pH 7 (T = 333, 343 K) were repeated using K_2CO_3 instead of Na_2CO_3 in the aging medium (Table 2-1, ID 9,10). As expected, no sodium zinc carbonate and no other by-phases were found in contrast to the aging experiment with Na_2CO_3 under the same conditions. This observation highlights the unexpected influence of the alkali metal counter ion on the chemistry of aging under these conditions. The crystallization of zincian malachite in the absence of Na^+ was strongly delayed (Figure 2-6a,d cf. ID 4 and ID 9 F; Table 2-1, cf. ID 4, 8 and 9, 10). With exception of the first two patterns, there was no significant down-shift in $d_{20\bar{1}}$ indicating the absence of any other transient Zn-storage phase (Figure 2-6h). Surprisingly, although the transient formation of sodium zinc carbonate was suppressed, the final $d_{20\bar{1}}$ values of the resulting zincian malachite phase were comparable to those obtained with Na_2CO_3. The crystallization of (Cu-rich) malachite can happen faster if a transient storage phase for Zn can form, but the final degree of Zn incorporation in zincian malachite is similar at a given pH value. This indicates that Zn incorporation is rather determined by the acidity of the aging medium, i.e. by thermodynamics, than affected by the transient segregation chemistry, i.e. by kinetics. It is the availability of H^+, which seems to promote or limit the zinc incorporation into zincian malachite and balances the ratio of Zn deposited into the Zn-sink phase Zn↓ (Figure 2-8) via the one or the other pathway. The lower limit of the pH value is, however, given by the partial dissolution of Zn^{2+} in acidic environment leading to incomplete solidification[36] or leaching and unfavorable Zn distribution (cf. section 2.3.4).

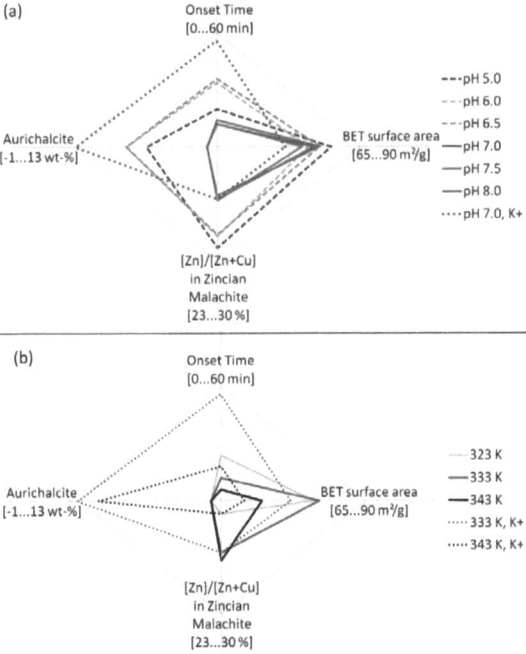

Figure 2-9: Radar plots illustrating the influence of the aging parameters pH value (a, at T = 333 K) and temperature (b, at pH = 7) on the aging reaction and the properties of the resulting zincian malachite material. Variation of pH leads to different aging mechanisms below pH 6.5 (a, broken lines) and above pH 7 (a, full lines) and affects the Zn content of the catalyst precursor. Temperature leads to a gradual change of the aging kinetics (b, full lines). Substitution of Na^+ by K^+ in the aging solution at the same temperature (a, dotted lines) has a pronounced effect on the aging reaction and the phase composition, but not on the Zn content of the zincian malachite precursor. As a function of temperature, a variation of the Zn content is observed with Na^+ and K^+.

2.4 Conclusion

The aging process of mixed Cu,Zn hydroxycarbonate precursors was decoupled from the precipitation and studied independently using *in-situ* EDXRD and *in-situ* UV-Vis spectroscopy. Crystalline zincian malachite, the desired precursor phase for Cu/ZnO catalysts, was successfully formed from the amorphous starting material in all experiments by aging in solutions with a composition near to the mother liquor under controlled co-precipitation conditions. As a function of different aging conditions, a variation of the Zn content in zincian malachite between ca. 24 and 29% was observed despite the same nominal Zn-content in the starting material of 30% indicating that a varying fraction of Zn was present in an undetected phase "Zn↓" acting as a sink for Zn. Two mechanisms to approach the maximal Zn incorporation into the zincian malachite catalyst precursor were observed: by direct co-condensation of Cu^{2+} and Zn^{2+} into Zn-rich malachite, or by first simultaneous crystallization of Cu-rich malachite and a transient Zn-storage phase, which in course of aging re-dissolved and allows for later Zn-enrichment of malachite. The latter mechanism is favored at pH $\geq$ 7 in the presence of Na^+ leading to crystallization of sodium zinc carbonate as Zn-storage phase. The former mechanism was observed at $5 \geq pH \geq 6.5$ and yields a higher Zn-incorporation into zincian malachite. The radar plots shown in Figure 2-9 summarize the effects of pH, temperature and alkali cation on the aging process. It can be seen that variation in pH changes the aging mechanism, while variation of temperature (at pH 7) leads to gradual changes. Thus, the acidity of the aging medium was identified as the most critical synthesis parameter to determine the final Zn-content in zincian malachite. Interestingly, Zn incorporation is independent of the crystallization mechanism. Even in the absence of Na^+, suppressing the transient crystallization of the sodium zinc carbonate storage phase, a lower degree of Zn incorporation was observed in the final sample at pH 7, although the reaction was following the direct co-condensation mechanism. The effect of individual synthesis parameters like temperature or acidity during catalyst preparation can be better rationalized on basis of the complex chemistry of precursor aging. They should be optimized to give a low amount of Zn↓ and a maximal Zn-substitution in malachite approaching the nominal Cu:Zn ratio of the synthesis.

Acknowledgement

This paper has emerged from a joint research project "Next generation methanol synthesis catalysts", which was funded by the German Federal Ministry of Education and Research (BMBF, FKZ 01RI0529). We acknowledge Edith Kitzelmann, Achim Klein-Hofmann, Gisela Lorenz, and Doreen Steffen for their substantial support in the lab, Elena Antonova, Jing Wang and Wolfgang Bensch for support with the EDXRD measurements, and HASYLAB (Hamburg, Germany) for allocation of beamtime. Martin Muhler and Stefan Kaluza are acknowledged for fruitful discussions. Robert Schlögl is greatly acknowledged for his continuous support.

2.5 References

[1] K. de Jong, *Synthesis of Solid Catalysts*, Wiley VCH., Weinheim, **2009**.
[2] M. Kurtz, N. Bauer, C. Buscher, H. Wilmer, O. Hinrichsen, R. Becker, S. Rabe, K. Merz, M. Driess, R. A. Fischer, M. Muhler, *Catal. Lett.* **2004**, *92*, 49-52.
[3] C. Baltes, S. Vukojevic, F. Schüth, *J. Catal.* **2008**, *258*, 334-344.
[4] B. Bems, M. Schur, A. Dassenoy, H. Junkes, D. Herein, R. Schlögl, *Chem. Eur. J.* **2003**, *9*, 2039-2052.
[5] I. Kasatkin, P. Kurr, B. Kniep, A. Trunschke, R. Schlögl, *Angew. Chem. Int. Edit.* **2007**, *46*, 7324-7327.
[6] R. G. Herman, K. Klier, G. W. Simmons, B. P. Finn, J. B. Bulko, T. P. Kobylinski, *J. Catal.* **1979**, *56*, 407-429.
[7] J. C. Frost, *Nature* **1988**, *334*, 577-580.
[8] V. Ponec, *Surf. Sci.* **1992**, *272*, 111-117.
[9] Y. Kanai, T. Watanabe, T. Fujitani, M. Saito, J. Nakamura, T. Uchijima, *Catal. Lett.* **1994**, *27*, 67-78.
[10] N. Y. Topsoe, H. Topsoe, *Top. Catal.* **1999**, *8*, 267-270.
[11] J. D. Grunwaldt, A. M. Molenbroek, N. Y. Topsoe, H. Topsoe, B. S. Clausen, *J. Catal.* **2000**, *194*, 452-460.
[12] T. Ressler, B. L. Kniep, I. Kasatkin, R. Schlögl, *Angew. Chem. Int. Ed.* **2005**, *44*, 4704-4707.
[13] M. Behrens, F. Studt, I. Kasatkin, S. Kühl, M. Hävecker, F. Abild-Pedersen, S. Zander, F. Girgsdies, P. Kurr, B.-L. Kniep, M. Tovar, R. W. Fischer, J. K. Nørskov, R. Schlögl, *Science* **2012**, *336*, 893-897.
[14] D. Waller, D. Stirling, F. S. Stone, M. S. Spencer, *Faraday Discuss.* **1989**, *87*, 107-120.
[15] S. Fujita, A. M. Satriyo, G. C. Shen, N. Takezawa, *Catal. Lett.* **1995**, *34*, 85-92.
[16] D. M. Whittle, A. A. Mirzaei, J. S. J. Hargreaves, R. W. Joyner, C. J. Kiely, S. H. Taylor, G. J. Hutchings, *Phys. Chem. Chem. Phys.* **2002**, *4*, 5915-5920.
[17] B. L. Kniep, T. Ressler, A. Rabis, F. Girgsdies, M. Baenitz, F. Steglich, R. Schlögl, *Angew. Chem. Int. Edit.* **2004**, *43*, 112-115.
[18] M. Behrens, *J. Catal.* **2009**, *267*, 24-29.
[19] A. M. Pollard, M. S. Spencer, R. G. Thomas, P. A. Williams, J. Holt, J. R. Jennings, *Appl. Catal. A* **1992**, *85*, 1-11.
[20] M. Behrens, F. Girgsdies, *Z. Anorg. Allg. Chem.* **2010**, *636*, 919-927.
[21] M. Behrens, F. Girgsdies, A. Trunschke, R. Schlögl, *Eur. J. Inorg. Chem.* **2009**, 1347-1357.
[22] R. Kiebach, N. Pienack, M. E. Ordolff, F. Studt, W. Bensch, *Chem. Mater.* **2006**, *18*, 1196-1205.
[23] L. Engelke, M. Schaefer, F. Porsch, W. Bensch, *Eur. J. Inorg. Chem.* **2003**, 506-513.
[24] N. Pienack, W. Bensch, *Angew. Chem. Int. Edit.* **2011**, *50*, 2014-2034.
[25] R. I. Walton, D. O'Hare, *Chem. Comm.* **2000**, 2283-2291.
[26] A. Michailovski, J. D. Grunwaldt, A. Baiker, R. Kiebach, W. Bensch, G. R. Patzke, *Angew. Chem. Int. Edit.* **2005**, *44*, 5643-5647.
[27] M. Behrens, R. Kiebach, J. Ophey, O. Riemenschneider, W. Bensch, *Chem. Eur. J.* **2006**, *12*, 6348-6355.
[28] F. Porsch, 3.155 ed., RTI GmbH, Paderborn, Germany, **2004**.
[29] L. Engelke, M. Schaefer, M. Schur, W. Bensch, *Chem. Mater.* **2001**, *13*, 1383-1390.
[30] A. Coelho, 4.2 ed., Bruker AXS GmbH, Karlsruhe, Germany, **2003-2009**.
[31] T. E. Gier, X. H. Bu, S. L. Wang, G. D. Stucky, *J. Am. Chem. Soc.* **1996**, *118*, 3039-3040.
[32] F. Zigan, W. Joswig, H. D. Schuster, S. A. Mason, *Z. Kristallogr.* **1977**, *145*, 412-426.

[33] S. Kaluza, M. Schroter, R. d'Alnoncourt, T. Reinecke, M. Muhler, *Advanced Functional Materials* **2008**, *18*, 3670-3677.
[34] H. Jung, D.-R. Yang, O.-S. Joo, K.-D. Jung, *B. Kor. Chem. Soc.* **2010**, *31*, 1241-1246.
[35] S. Klokishner, M. Behrens, O. Reu, G. Tzolova-Müller, F. Girgsdies, A. Trunschke, R. Schlögl, *J. Phys. Chem. A* **2011**, *115*, 9954-9968.
[36] J. L. Li, T. Inui, *Appl. Catal. A* **1996**, *137*, 105-117.
[37] M. Behrens, D. Brennecke, F. Girgsdies, S. Kißner, A. Trunschke, N. Nasrudin, S. Zakaria, N. F. Idris, S. B. Abd Hamid, B. Kniep, R. Fischer, W. Busser, M. Muhler, R. Schlögl, *Appl. Catal. A* **2011**, *392*, 93-102.

Supplementary Information

Table S2-1: Aging parameters and internal sample numbers. The sample ID 0 refers to the unaged precursor.

ID	pH	T [K]	A^+ in A_2CO_3	Internal sample number
0	-	-	-	7005
1	5	333	Na^+	7037
2	6	333	Na^+	7057
3	6.5	333	Na^+	7064
4	7	333	Na^+	7045
5	7.5	333	Na^+	7036
6	8	333	Na^+	7038
7	7	323	Na^+	7044
8	7	343	Na^+	7043
9	7	333	K^+	7063
10	7	343	K^+	7058

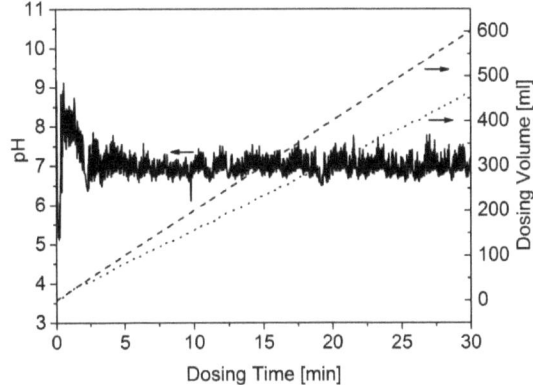

Figure S2-1: Evolution of pH (black curve) with added Cu,Zn solution (dashed) and Na_2CO_3 solution (dotted) during co-precipitation of the Cu,Zn (70:30) precursor in the continuous process. The slurry was continuously fed into a spray-dryer.

Chapter 2: Supplementary Information

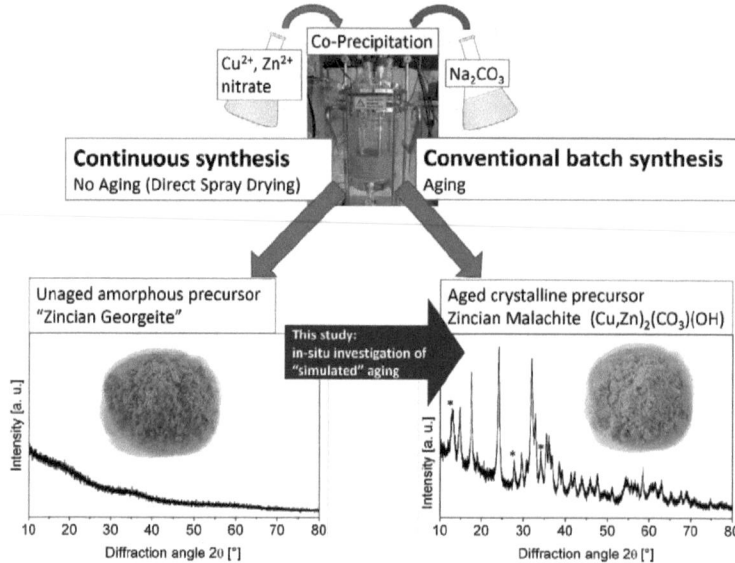

Figure S2-2: Schematic representation for the T- and pH-controlled precursor preparation from aqueous solutions using an automated laboratory-reactor. The co-precipitation and aging stages (right hand side) were decoupled by continuously removing the "unaged" precursor (left hand side) and subsequent aging studies at varying conditions with the same starting material (large arrow). The marked reflections in the lower right hand corner XRD pattern refer to the aurichalcite by-phase, $(Cu,Zn)_5(CO_3)_2(OH)_6$. All other reflection are due to zincian malachite $(Cu,Zn)_2(OH)_2CO_3$.

Chapter 2: Supplementary Information

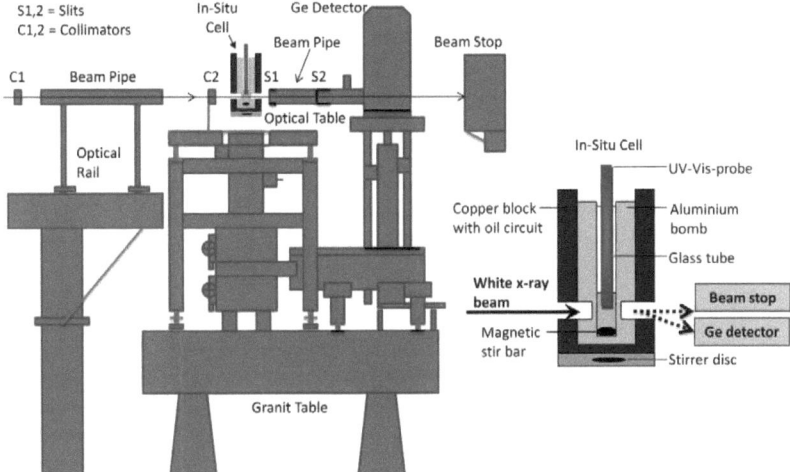

Figure S2-3: Detailed experimental setup of *in-situ* EDXRD reaction cell at the F3 beamline at HASYLAB, Hamburg, Germany.

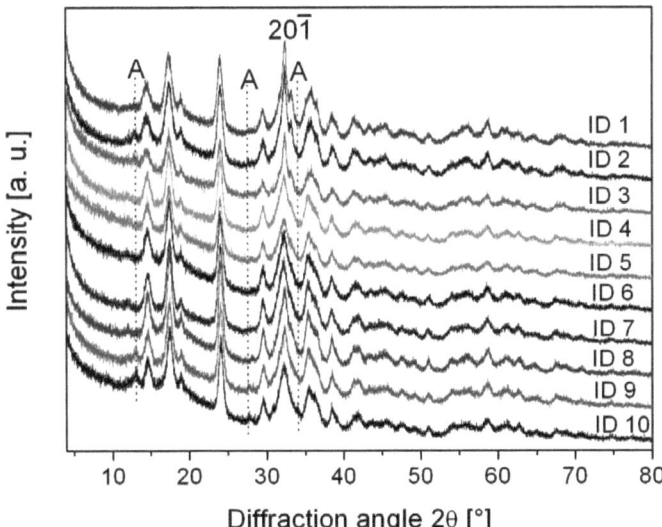

Figure S2-4: *Ex-situ* XRD patterns of all recovered sample (for labeling see Table 2-1). The 20-1 reflection of malachite and the characteristic peaks of aurichalcite are marked. The labeling refers to the entry number of Table 2-1 in the main article.

39

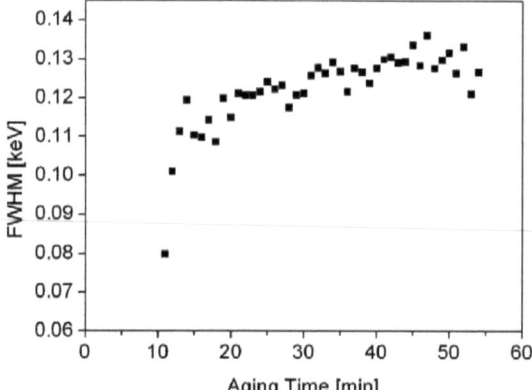

Figure S2-5: Evolution of the FWHM of the 20-1 reflection of zincian malachite from the EDXRD spectra during simulated aging at pH 7, 323 K (ID 7). In all aging experiment where reflections of the sodium zinc intermediate were detected, the FWHM of the 20-1 reflection of zincian malachite did not decrease during disappearance of the sodium zinc carbonate phase. Thus, the shift of the 20-1 reflection seems not to be an effect of overlapping peaks from both phases but rather of Zn incorporation into the zincian malachite phase.

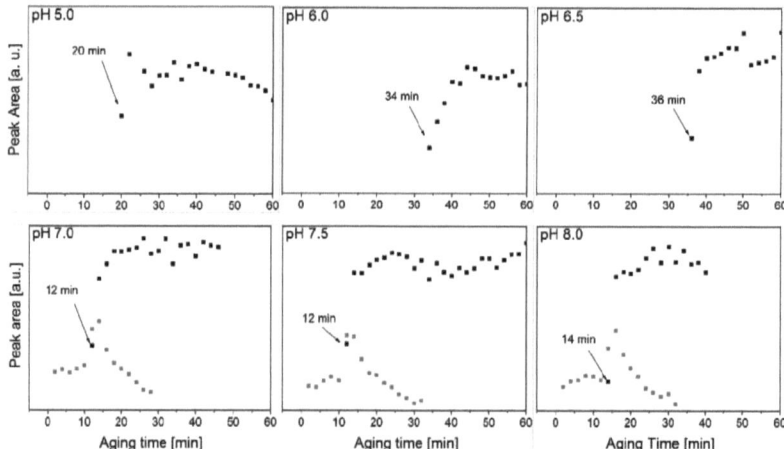

Figure S2-6: Integral intensity of selected EDXRD peaks of detected phases vs. aging time in Na_2CO_3 at T = 333 K at different pH-values. Zincian malachite $(Cu,Zn)_2(OH)_2(CO_3)$ is represented by the 20-1 reflection (black), sodium zinc carbonate $Na_2Zn_3(CO_3)_4 \cdot 3H_2O$ is represented by the 222 peak (grey).

Chapter 3: Correlations between Preparation and Microstructure of Cu/ZnO Catalysts for Methanol Synthesis – Influence of the pH value during Synthesis of Cu,Zn Hydroxy Carbonates

Stefan Zander, Igor Kasatkin, Gisela Weinberg, Patrick Kurr, Benjamin Kniep, Malte Behrens.

Abstract

The co-precipitation of mixed Cu,Zn hydroxide carbonate precursors is the first step during preparation of Cu/ZnO catalysts for methanol synthesis. The pH value influences the precursor chemistry because it controls the precipitation behavior and also the subsequent aging process. Application of different pH values in the range of pH 6.0-9.0 were chosen based on previous results. To check for the reproducibility of the kinetically controlled co-precipitation process, two sample series were prepared under identical conditions. Use of pH values ≥ 6.5 led to higher phase fraction of zincian malachite at the expense of the undesired Zn-rich by-phase aurichalcite. As a consequence, more Zn was inserted into zincian malachite which was verified from the position of the $20\bar{1}$ reflection by Rietveld refinement of the XRD patterns of the precursors. Samples prepared at pH 7.5 and higher showed a split up signal of the $20\bar{1}$ reflection indicating two different zincian malachite phases with different Zn substitution. Samples prepared at 6.0 ≤ pH ≤ 7.0 exhibited a better homogeneity of the Zn distribution within the zincian malachite. After calcination, samples prepared at pH 6.0 showed the largest CuO domain size. Reduced Cu/ZnO samples exhibited Cu surface areas in the range of 18 to 20 $m^2\ g^{-1}$. Only samples prepared at pH 8.5 showed a larger Cu surface area of around 25 $m^2\ g^{-1}$. No simple correlation was found between the Cu surface area and any microstructural characteristics of the samples. It is expected that the activity of the catalysts scales linearly with the Cu surface area.

Chapter 3: Correlations between Preparation and Microstructure of Cu/ZnO Catalysts for Methanol Synthesis – Influence of the pH value during Synthesis of Cu,Zn Hydroxy Carbonates

3.1 Introduction

Cu/ZnO/(Al$_2$O$_3$) catalysts are of major industrial interest as they have been successfully applied in methanol synthesis for over 40 years. These catalysts are also used in methanol steam reforming and water-gas-shift reaction. The synthesis route for preparing Cu/ZnO/(Al$_2$O$_3$) catalysts follows a multi-step procedure including temperature- and pH-controlled co-precipitation of aqueous Cu,Zn,Al nitrate solution with sodium carbonate solution as precipitating agent. Subsequently, the precipitate is aged in the mother liquor, filtrated, washed and dried to obtain the mixed metal hydroxy carbonate precursor phase. After forming the metal oxides by calcination, reduction of CuO to metallic copper yields the active catalyst [1].

All parameters applied in every single step of the catalyst preparation can influence the bulk and surface structure and therewith the characteristics and activity of the resulting catalyst. This phenomenon is also called the "chemical memory" and relates to the influence of early stage parameters on the characteristics of the precursor phase and, finally, on the microstructure and activity of the resulting catalyst [2-3]. Accordingly, the precipitation and aging conditions were shown to have a significant impact on the final catalyst [1, 4-13]. Regarding the activity of the resulting catalyst, co-precipitation of Cu,Zn,Al systems turned out to be most successful applying pH 7 and a temperature of 343 K [10-11]. Baltes et al. [11] reported that variation of pH values and temperature during the first step of the catalyst preparation for the industrially relevant ternary system (Cu:Zn:Al = 60:30:10) yielded differences in Cu surface areas which were found to be linearly correlated with the catalyst's activity in methanol synthesis. While their study was rather focused on the physico-chemical properties of the calcined and reduced samples, we herein report a systematic study of pH variation on the properties of the hydroxy carbonate precursors.

To reduce the complexity of ternary Cu,Zn,Al systems, often binary Cu,Zn functional model systems are applied. Bems et al. investigated phase formation and thermal decomposition of precursors with different Cu:Zn ratios and found Cu$_2$(OH)$_2$CO$_3$ (malachite) for pure Cu samples, (Cu$_{1-x}$Zn$_x$)$_2$(OH)$_2$CO$_3$ (zincian malachite) with x < 0.3 and (Cu$_{1-y}$Zn$_y$)$_5$(OH)$_6$(CO$_3$)$_2$ (aurichalcite) with y > 0.5 as the predominant precursor phases. In the XRD patterns, they observed a shift of the 20$\bar{1}$ reflection of zincian malachite with increasing Zn content in this phase [3], which also was reported earlier by Porta et al. [14]. This shift can be correlated to the accessible copper surface area of the final catalyst and to a first approximation also to the catalytic activity [15]. Thus, the critical role of the precursor chemistry has been emphasized as follows: To obtain highly active catalysts, zincian malachite is the relevant precursor phase because it combines effects of particle morphology and Zn content better than aurichalcite. As a result, a higher Cu surface area is obtained after meso- and nanostructuring during catalyst synthesis. Nevertheless, the presence of aurichalcite indicates that the maximal Zn incorporation into the zincian malachite phase is reached.

The critical step for precursor phase formation is the aging of the precipitate upon which it crystallizes. In a recent in-situ study of the aging process, we identified pH of 6.5 and lower and temperatures of 333-343 K to result in the highest Zn incorporation in the zincian malachite phase for decoupled precipitation and aging in a down-scaled synthesis of a few milligrams [13]. The aim of the present study was to elucidate the impact of the pH value during precipitation for larger batch synthesis using a 2 L-volume reactor with yields in the 0.1 kg scale. Furthermore, analysis of calcined and reduced samples should deliver correlations between synthesis pH and microstructure, in particular test the proposed correlation of the position of the $20\bar{1}$ reflection of zincian malachite with the accessible copper surface area which is closely related to the catalytic activity in methanol synthesis.

3.2 Experimental

3.2.1 Sample Preparation

Metal hydroxy carbonate precursors with fixed Cu:Zn ratio (70:30) were synthesized by co-precipitation from acidic Cu,Zn nitrate solution and Na_2CO_3 solution as basic precipitating agent in an automated lab reactor (LabMax, Mettler Toledo) under controlled conditions like dosing, stirring, temperature (338 K) and constant pH value. While all other parameters were fixed, the pH value was varied in different experiments between 6.0 and 9.0 in steps of 0.5 pH units. The range of pH was chosen because pH values lower than 6.0 lead to incomplete Zn precipitation, pH values higher than 9.0 to oxolation of the precipitates to form CuO, which is undesired in this stage of preparation [9]. The co-precipitate slurry was aged until a transient pH drop [3, 15] was observed and stirred for additional 30 min. Then, the slurry was filtered, the aged precipitate was washed several times with water and spraydried (Niro minor mobile, T_{inlet} = 473 K, T_{outlet} = 373 K). Calcination was carried out in static air at 603 K (2 K min^{-1}) for 3 h. The high Cu:Zn ratio of 70:30 is typically applied in industrial catalyst preparation and aims at a maximal incorporation of Zn into the zincian malachite precursor phase [15]. The designations of the samples are given in Table S3-1.

3.2.2 Characterization

X-ray fluorescence spectroscopy (XRF) of the calcined samples was performed with powders or after glassing with $Li_2B_4O_7$ in a Bruker S4 Pioneer X-ray spectrometer.

X-ray diffraction (XRD) was applied to the catalyst precursors and calcined samples. The samples were measured on a STOE STADI P transmission diffractometer equipped with a primary focusing Ge monochromator (Cu $K_{\alpha 1}$ radiation) and a linear position sensitive detector (moving mode, step size 0.1 °, counting time 10 s/step, resolution 0.01 °, total accumulation time 634 s). The samples were mounted in the form of a clamped sandwich of small amounts of powder fixed with a small amount of grease between two layers of thin polyacetate film. The phase composition was determined by full pattern refinement in the 2θ range 4-80 ° according to the Rietveld method using the TOPAS software [16] and crystal structure data from the ICSD database.

Specific surface areas were determined by N_2 physisorption in a Quantachrome Autosorb-6 machine after degassing the samples at 353 K for 2 h. Isotherms were recorded at liquid nitrogen temperature and evaluated according to the BET method. The recorded isotherms of all precursors and calcined samples featured the characteristics of mesoporous substances (not shown).

Thermogravimetric experiments (TGMS) were done on a NETZSCH Jupiter thermobalance in flowing air at a heating rate of 2 K min^{-1}. The gas evolution was measured with a quadrupole mass spectrometer (Pfeiffer Vacuum, Omnistar).

Scanning electron microscopy (SEM) images were taken in a Hitachi S-4800 field emission gun (FEG) system. Transmission electron microscopy (TEM) was performed with a Philips CM200FEG microscope operated at 200 kV and equipped with an EDX spectrometer. For TEM investigation, the samples were reduced up to a temperature of 523 K and transferred to the microscope in inert atmosphere. The coefficient of spherical aberration was Cs = 1.35 mm and the information limit was better than 0.18 nm. High-resolution images with a pixel size of 0.016 nm were acquired at the magnification of 1083000x with a CCD camera, and selected areas were processed to obtain power spectra (square of the Fourier transform of the image), which were used for measuring interplanar distances and angles (accuracy ± 1% and ± 0.5 deg, correspondingly) for phase identification. Projected areas have been measured and equivalent diameters calculated for 1500-3000 Cu particles in each sample. In all cases the values of standard error of the mean diameter were ≤ 0.1 nm. Frequency distributions of the particle sizes fitted well to Lognormal functions. EDX analyses were performed for 5-15 larger aggregates containing at least several hundred particles in each sample.

Temperature programmed reduction (TPR) was performed with around 40 mg of each sample in a glass reactor, fixed by means of quartz wool plugs. The reduction was carried out in a CE instruments TPDRO 1100 machine with 80 mL min^{-1} 5% H_2 in Ar up to a temperature of 623 K (6 K min^{-1}). The K-values according to Monti and Baiker [17] were 100-120 s, the P-values according to Caballero [18] 10-12 K. The reduction progress was followed with an internal thermal conductivity detector. Analysis was performed with regard to the temperature with the highest H_2 consumption (T_{max}) and the total H_2 consumption with respect to the CuO content in the sample (compared with a pure CuO reference). In the following, the term "reducibility" is used for the latter feature. The CuO content of the samples was derived from XRF data with the assumption that only CuO and ZnO were present and under neglect of residual carbonate species (see below).

The copper surface area was determined by applying N_2O reactive frontal chromatography (N_2O-RFC) based on the method proposed by Chinchen et al. [19]. Around 100 mg of a sieve fraction (100-200 µm) of each sample were placed in a stainless steel U-tube reactor and fixed by means of quartz wool plugs. The prior reduction was carried out in the same device and conditions as for TPR, but only up to a temperature of 523 K and with a holding time of 30 min. The reduction progress was additionally followed with a quadrupole mass spectrometer (Pfeiffer Vacuum, Omnistar). After cooling down to 303 K, the catalyst has been flushed for 30 min in pure Ar and 15 min in pure He in order to achieve an adsorbate-free reduced catalyst surface. N_2O-RFC was performed with 10 mL min^{-1} 1% N_2O in He, at which the N_2O reacts quantitatively with the Cu surface atoms forming gas-phase N_2. The specific Cu metal surface area has been calculated from the formed amount of N_2 using a value of $1.47*10^{19}$ atoms per m^2 for the mean Cu surface atom density. The error of the specific Cu surface area is about ± 1 m^2 g^{-1}.

3.3 Results and Discussion

Two identical sets of samples were prepared to check for the reproducibility of our preparation and estimate the role of possible "batch effects" on the results. All characterizations were performed for both sets of samples unless otherwise noted. The designation of the binary Cu,Zn samples was chosen with regard to pH value of the precipitation, respectively. CZ6.0 means that the sample(s) were prepared at pH 6.0. For reasons of visualization, the shown values in the plots and tables include values of the original and the reproduction experiment as well as the average values. The endings of the given error bars represent the real values obtained for the two samples for each precipitation pH value and give information about the reproducibility of the sample preparation.

3.3.1 Precipitation and aging

A typical evolution of pH value and turbidity during precursor preparation is shown in Figure 3-1 and is similar to the data described in ref. [15] in detail. The pH value is kept at a constant value (here: 8.5) during simultaneous dosing of metal solution and precipitation agent. After the end of dosing ($t_{aging} = 0$), the free aging begins and is characterized by an uncontrolled evolution of the pH value. The pH drop ($t_{aging} = 14$ min) is accompanied by an increase in the turbidity and a change of the color of the slurry from blue to green. The initially amorphous precipitate is transformed into a crystalline product [15].

Having a closer look to the recorded pH traces of all samples during the aging of the freshly precipitated solids (after dosing) some trends can be observed within the whole pH series. First, we inspect the duration until the local minimum of the pH drop (Figure 3-2a). For CZ6.0 the averaged time amounts 142 minutes and decreases down to 12 minutes for CZ8.0. After this minimum, the value increases again up to 33 minutes (CZ9.0). An earlier pH drop with increasing preparation pH values was also observed in our former study where we decoupled precipitation and aging and investigated the aging process at different pH values from 5.0 to 8.0 [13]. The error bars, i.e. the reproducibility within two experiments, of the obtained values in Figure 3-2a are remarkably low. Only CZ7.0 showed large error bars indicating differences in the crystallization kinetics between the two runs.

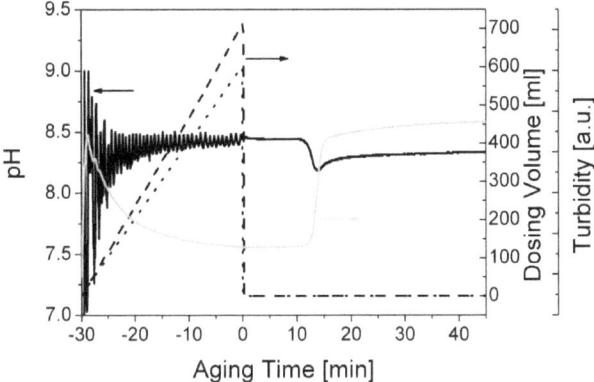

Figure 3-1: Evolution of pH (black curve), turbidity (grey curve) during pH-controlled (pH 8.5) dosing of acidic Cu,Zn solution (dotted curve) and basic Na₂CO₃ solution (dashed curve) and subsequent free aging of a Cu,Zn (70:30) precursor.

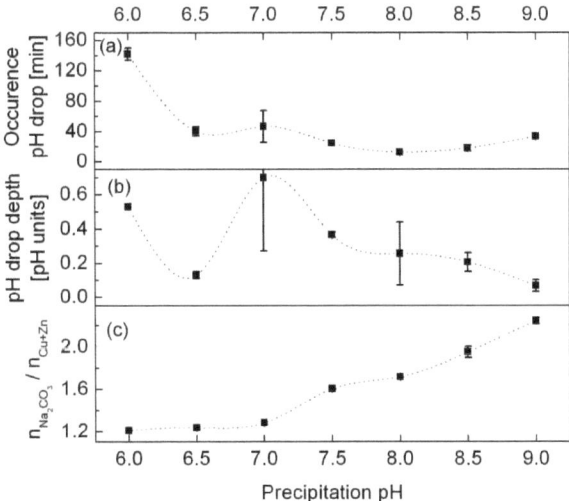

Figure 3-2: Results from the recorded data during precipitation and aging in dependence on the precipitation pH value: time of the occurrence of the pH drop (a), depth of the pH drop (b), ratio of dosed moles of Na₂CO₃ and metals (Cu+Zn) (c). Shown values are representing average values, endings of the error bars represent the real values obtained for the two samples for each precipitation pH value.

As a second feature, the depth of the pH drop was evaluated as a hint, how pronounced the chemical transformations during the pH drop are (Figure 3-2b). Again, the largest deviation between the two experiments was observed for the synthesis conducted at pH 7.0. With exception of pH 6 and the one experiment at pH 7, similar values without a general trend were observed.

Another characteristic is the ratio of added CO_3^{2-} and metal (Cu^{2+} + Zn^{2+}) ions. While the reproducibility is good for all experiments, the amount of CO_3^{2-} ions in the reaction mixture is generally higher for higher precipitation pH values. This is expected, because Na_2CO_3 is used to regulate the pH value during precipitation whereas the amount of added Cu^{2+} and Zn^{2+} ions stays the same for every preparation (Figure 3-2c). Nevertheless, two different regions can be identified. In the precipitation range of pH 6.0 to pH 7.0, the consumption of Na_2CO_3 is relatively constant, whereas a linear increase can be observed starting at pH 7.5.

The kinetics of crystallization of the hydroxide carbonate precursors will depend on the CO_3^{2-} and OH^- concentrations and this dependency is reflected in changes in the time period until occurrence of the pH drop and its depth. The details and elementary steps of this event are not easily probed experimentally and no obvious correlation can be established between the observed changes in pH and specific chemical reactions. It is known that the pH drop can be assigned to modifications in the phase composition by dissolution and re-precipitation leading to crystallization of the initially amorphous precipitate zincian georgeite to crystalline zincian malachite and aurichalcite [3-4, 8]. Effectively, such a reaction will be related to a transient exchange of anions (and cations) between the solid precursor and the solution leading to the observed characteristic pH traces.

To check for a complete precipitation of all Cu,Zn species, the Cu,Zn oxides (calcined samples) were subjected to elemental analysis by XRF. The experimental values (Table S3-1) were in good agreement with the nominal Zn content of 30% (x_{Cu} + x_{Zn} = 100%) from the initial Cu,Zn solution, with two exceptions: For CZ7.5, there was a low, but significant increase of the Zn content. It should be noted, that this behavior was reproducible but was not found in the results of Baltes et al.[11] for the Cu,Zn,Al system. CZ9.0 showed a loss of Zn. However, regarding the large error in this case, reproducibility seems to be limited at this pH value. These minor variations are tentatively related to formation of soluble species at the respective pH or by loss of Cu- or Zn-enriched materials during recovering the samples, e.g. due to preferred sticking to the walls of the glass reactor.

3.3.2 Precursor and calcined materials

Figure 3-3 shows the XRD patterns of one set of precursors which is also qualitatively representing the set of reproduced samples. As expected, aurichalcite and zincian malachite were identified as the only crystalline phases for all precursor samples.

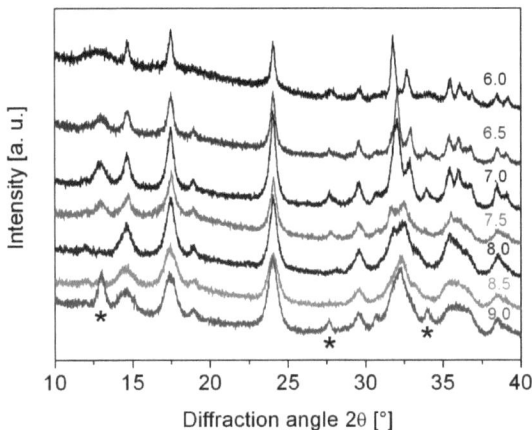

Figure 3-3: XRD patterns of one selected precursor sample for each preparation pH value. The well-resolved peaks of aurichalcite are marked with a star. All other reflections can be assigned to zincian malachite or to an overlap of peaks from both phases.

XRD patterns of all precursor samples and calcined samples were subjected to Rietveld refinement for phase identification and determination of composition and crystallite domain sizes. As a representative example, the refinement result of a sample prepared at pH 7.5 is depicted in Figure 3-4a. It is noted that the accurate determination of the exact weight fractions in the phase mixture is difficult due to the low amounts of aurichalcite, the generally low crystallinity and the high noise of the XRD patterns. The absolute values of individual samples depend on the fitting constraints and have to be compared with care. However, the general trend seen within the series of samples are regarded as reliable, as the used fitting constraints have been equal for all samples. Depending on the preparation pH, some fits in the region between 31 and 33 ° 2θ were dissatisfying and the corresponding $20\bar{1}$ reflection signal of zincian malachite appeared to be split up into two peaks (inset, Figure 3-4a). Since the position of the $20\bar{1}$ reflection is a measure for the degree of Zn incorporation in the zincian malachite phase, apparently two differently Zn-substituted zincian malachite phases were present. This double peak can already be seen in Figure 3-3 for CZ7.5 and CZ8.0. A Rietveld refinement using two independent zincian malachite phases (and aurichalcite) led to a better agreement of the experimental data with the simulated data in the mentioned region (Figure 3-4b). As reported in ref. [13], where the aging process was investigated in detail, two different mechanisms can lead to the formation of zincian malachite: 1) direct co-condensation of Cu^{2+} and Zn^{2+} into a Zn-rich malachite; 2) first simultaneous crystallization of Cu-rich malachite and a transient Zn-storage phase, which in course of aging re-dissolves and allows for later Zn-enrichment of malachite. The former mechanism was observed at $5 \leq pH \leq 6.5$, resulting in a higher Zn-incorporation into zincian malachite. The latter mechanism was found to be favored at $7 \leq pH \leq 8$ in the presence of Na^+ leading to crystallization of sodium zinc carbonate as Zn-storage phase. The observation of two different zincian malachite phases in the precursor material recovered at pH 7.5 and pH 8.0 indicated that in our experiments both mechanisms may operate simultaneously. The Zn-richer component was presumably formed via mechanism 1), while the Zn-poorer fraction of zincian malachite has formed via mechanism 2).

Chapter 3: Correlations between Preparation and Microstructure of Cu/ZnO Catalysts for Methanol Synthesis – Influence of the pH value during Synthesis of Cu,Zn Hydroxy Carbonates

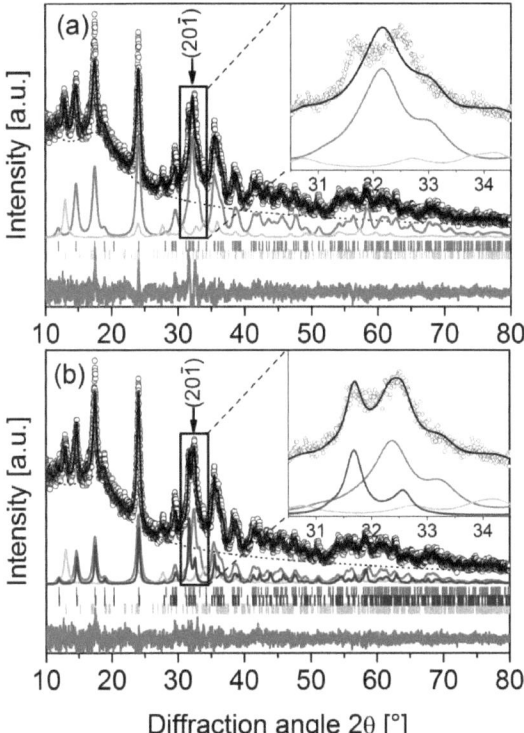

Figure 3-4: Rietveld refinement for the XRD pattern of a precursor sample prepared at pH 7.5: With one zincian malachite phase (a); with two zincian malachite phases (b); experimental data (open circces), total calculated curve (black), background (dotted grey), difference curve (grey), calculated pattern zincian malachite 1 (grey), calculated pattern zincian malachite 2 (dark grey), calculated pattern aurichalcite (light grey). The thick marks indicate the positions of the Bragg reflections.

In Figure 3-5a (black symbols), the fractions of aurichalcite in the precursor samples are shown. CZ6.0 contained most aurichalcite (30 wt%) at the expense of zincian malachite. Precipitation pH values of 6.0 and higher led to a fraction of 5 to 10 wt% aurichalcite but no distinct minimum was detected. The averaged BET surface area values (Figure 3-5a, grey symbols) for precursors lay in the range of 49 m^2 g^{-1} (CZ7.5) to 127 m^2 g^{-1} (CZ6.0) again without an obvious trend with variation in synthesis pH value.

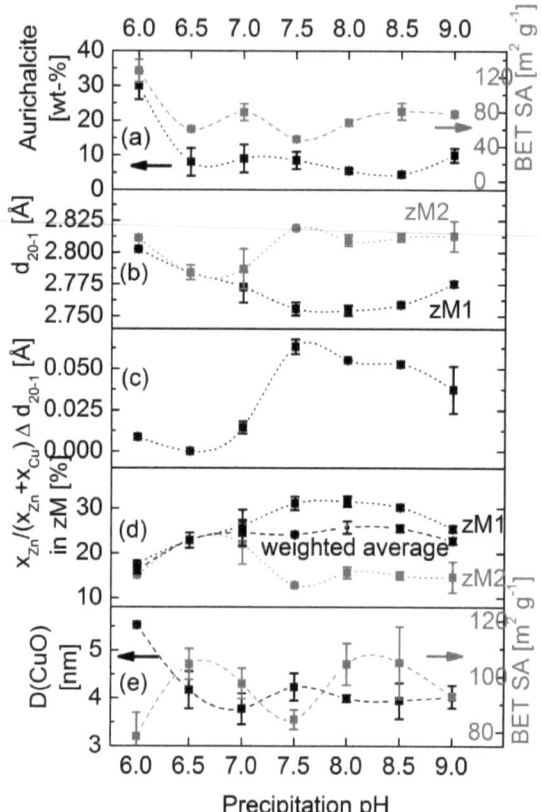

Figure 3-5: Results of XRD full pattern refinement and BET surface areas of Cu,Zn hydroxy carbonate precursors and calcined samples in dependence on the precipitation pH value. The zincian malachite phase "zM1" and "zM2" refer to the refinement with two different zincian malachite phases: weight fraction of the aurichalcite phase (black) and BET surface areas (grey) of the precursors (a); d-spacing of the $20\bar{1}$ reflection of zM1 (black) and zM2 (grey) (b); difference between the $d_{20\bar{1}}$ values of both zincian malachite phases zM1 and zM2 (c); Zn-content ($x_{Cu} + x_{Zn}$ = 100%) in zM1 (black), zM2 (grey) and overall weighted average Zn content in both zincian malachite phases (black dashed) (d); domain size of CuO (black) and BET surface areas (grey) of the calcined samples (e). Shown values are representing average values, endings of the error bars represent the real values obtained for the two samples for each precipitation pH value. Lines are guides for the eyes.

In Figure 3-5b, the $d_{20\bar{1}}$ values are shown. These were obtained resulting from full pattern refinement using two different zincian malachite phases for all samples, even those which can be satisfactorily fitted with only one zincian malachite phase, to allow comparability. The $d_{20\bar{1}}$ value can be used as a quantitative measure of the incorporation of non Jahn-Teller-distorted

Zn^{2+} into this phase and was first reported by Porta et al. [14]. Low $d_{20\bar{1}}$ values correspond to high incorporation. The $d_{20\bar{1}}$ value of pure malachite, $Cu_2(OH)_2CO_3$, is about 2.863 Å. The Zn-richer phase is labeled Zincian malachite 1 (zM1, black symbols) and exhibits a lower $d_{20\bar{1}}$ value than that of zincian malachite 2 (zM2, grey symbols). For CZ6.0 and CZ6.5 the $d_{20\bar{1}}$ values of zM1 and zM2 are almost identical confirming that indeed only one zincian malachite phase with a homogeneous Zn content is present. For pH values of 7.0 and higher, the two curves separate indicating inhomogeneous Zn distribution within the zincian malachite. The lowest $d_{20\bar{1}}$ values for zM1 were obtained for CZ7.5 and CZ8.0 (2.75 Å).

The difference of the both obtained $d_{20\bar{1}}$ values (zM1 and zM2) gives an insight into the homogeneity of the Zn distribution within the zincian malachite phases (Figure 3-5c). Homogeneous distribution, i.e. a single zincian malachite phase zM1 = zM2, was found for CZ6.0, where the difference is quite small, and for CZ6.5, where it is exactly zero. For CZ7.0, the difference is still quite small, but higher pH values lead to larger differences caused by inhomogeneous Zn distribution. Remarkably, the reproducibility of this feature is very good. The error bars are small and in some cases, reproduction delivers again the same values. Low reproducibility was again detected for CZ9.0. Inhomogeneity of the particle morphology and fluctuation of the local Cu:Zn ratio was also observed by SEM-EDX investigation on selected samples (for details see supporting information).

The $d_{20\bar{1}}$ value can be directly converted into the Zn content ($x_{Zn} + x_{Cu} = 100\%$) in the zincian malachite phase according to the correlation of the Cu,Zn system shown in [13, 20-21]. The results are plotted in Figure 3-5d and extreme values are obtained for zM2 (13%, CZ7.5, grey symbols) and zM1 (32%, CZ8.0, black symbols). It is noted that the latter value exceeds the proposed maximum of 28% synthetic zincian malachite samples prepared at pH 7.0 [21].

Including the phase fractions of zM1 and zM2, an overall weighted average Zn content in the malachite phases was calculated (Figure 3-5d; black symbols, dashed line). These averaged Zn contents are below the nominal Zn content of 30% used during synthesis and the XRF results. The remainder of Zn is assumed to be present in the aurichalcite by-phase. The lowest average value of 16% was found for CZ6.0, the sample that also shows the highest aurichalcite fraction. All other samples exhibited average Zn contents in the zincian malachite component between 23% (CZ6.5) and 26% (CZ8.0).

Chapter 3: Correlations between Preparation and Microstructure of Cu/ZnO Catalysts for Methanol Synthesis – Influence of the pH value during Synthesis of Cu,Zn Hydroxy Carbonates

To interpret the data obtained from XRD pattern refinement, it is important to recapitulate that zincian malachite is the relevant precursor phase to obtain highly active catalysts [15]. These results show that not only the average Zn incorporation into this catalyst precursor phase plays a decisive role, but that at higher precipitation pH also the homogeneity of the Zn distribution within the zincian malachite phase fraction has to be considered. Both features can influence the dispersion of CuO particles, separated by ZnO particles, after calcination. Higher Zn incorporation into the zincian malachite should increase the inter-dispersion of ZnO and CuO particles. A more homogeneous Zn distribution in the zincian malachite should cause a narrower statistical distribution of the CuO particle size.

Under these assumptions, the samples prepared at pH 6.0 can be estimated to result in poor catalysts, because they show large fractions of aurichalcite and therefore contain little Zn in the zincian malachite phase, although it is homogeneously distributed. The preparation at pH 6.5 and higher results in an increased Zn incorporation. Samples prepared at pH 6.5 and 7.0 seem to be most suitable because they additionally show a good homogeneity, in contrast to pH 7.5 and higher. In the latter case, the nanostructuring of the CuO after calcination will proceed probably not equally efficiently for both the zM1 and the zM2 phases, while there might be a compensating effect of the Zn-richer domains for the Zn-poorer fraction of the catalyst. With these considerations in mind, the calcined samples have been investigated to shed more light on this issue.

The process of calcination has been monitored by TGA-EGA and the results of this study are reported as supporting information. Due to the similar phase contrast of ZnO and CuO, statistical evaluation of TEM data of the calcined samples is not easy and here we use the domain size of the CuO crystallites as determined from XRD as a proxy for the particle size. "Domain (crystallite) size" values from XRD should not be mistaken for the real "particle size" as observed (in projection) in TEM images, which is usually higher because particles may consist of several domains. This has been shown in a detailed XRD and TEM study for reduced Cu/ZnO/Al_2O_3 catalysts by Kasatkin et al. [22]. However, in nanostructured samples, where the particle size is in the low nm-range and each particle contains several hundreds or thousands of unit cells, there usually is a clear relation between XRD and TEM size, because the number of domain boundaries in one particle is limited by geometrical reasons [23]. Typically a reliable correlation of XRD and TEM size is observed in such materials.

Rietveld refinement of the XRD patterns of the calcined samples (not shown) revealed CuO as the main phase and only small amounts (up to 2 wt%) of crystalline ZnO indicating that most of the ZnO is amorphous or finely dispersed. The domain size of CuO is shown in Figure 3-5e (black symbols). CZ6.0 exhibited the largest domain size (5.5 nm) as expected from the poor inter-dispersion of CuO and ZnO as a result of low Zn-incorporation into the zincian malachite phase. All other values are very similar around ≈ 4 nm and the differences are probably not significant with regard to the error bars. The difference between CZ6.0 and the other samples 5.5 and 4 nm seems to be small but a simple geometric calculation reveals that the surface area of spherical particles increases from 100% (5.5 nm) to 137.5% (4 nm) at a constant total mass. The BET surface areas of the calcined samples (Figure 3-5e, grey symbols) exhibited somewhat higher values compared to the precursors (Figure 3-5a, grey line) with the exception of CZ6.0, which cannot undergo efficient nanostructuring during calcinations due to a lack of Zn in the zincian malachite precursor phase. All other samples increased their surface area after thermal decomposition. This effect was most prominent for CZ6.5 and CZ7.5, where the surface area was increased by around 75%. The BET values are in a narrower range of 79 $m^2 g^{-1}$ (CZ6.0) to 105 $m^2 g^{-1}$ (CZ6.5, CZ8.0 and CZ8.5) compared to the precursor materials. They were in line with the double maximum around pH 6.5 and 8.5 found by Baltes et al. for ternary CuO/ZnO/Al$_2$O$_3$ samples (Fig.3 for T = 338 K ≈ 65 °C in ref. [11]). An inverse correlation is observed between BET surface areas and CuO domain size. This is expected because small CuO domains are a result of better nanostructuring which leads to higher BET surface areas.

To conclude this part, it was shown that the pH value of synthesis has only a low impact on the phase composition regarding zincian malachite and aurichalcite for pH ≥ 6.5 and on the Zn incorporation into the former phase. However, a pronounced effect on the Zn distribution in the zincian malachite precursor was observed. Inhomogeneous Zn distributions with a Zn-poorer and a Zn-richer zincian malachite component were observed for pH ≥ 7.5. The pH range typically applied for catalyst preparation of 6.5 – 7.0 yields homogeneous zincian malachite precursors with a high and uniform degree of Zn incorporation. An effect of the Zn incorporation on the ZnO-CuO inter-dispersion after calcinations and on the microstructural homogeneity of the resulting catalysts is expected and will be in the focus of the next section of this work.

3.3.3 Reduction and reduced samples

The reduction behavior of the calcined Cu,Zn oxide samples was investigated by temperature programmed reduction (TPR) in hydrogen and the results are reported as supporting information.

TEM investigation was done on one reduced sample on CZ6.5 and CZ9.0, representing the two precursor families with homogeneous and inhomogeneous Zn distribution in the zincian malachite precursor. CZ6.5 showed round shaped Cu particles (Figure 3-6a) and a mean Cu particle size of around 11.1 nm (Figure 3-6b). According to the local elemental composition from TEM-EDX at different spots, the Cu:Zn ratio was 69.8:30.2. The low standard deviation (2.0) shows, that the sample was quite homogeneous and the composition is in reasonable agreement with the findings from XRF of the calcined sample.

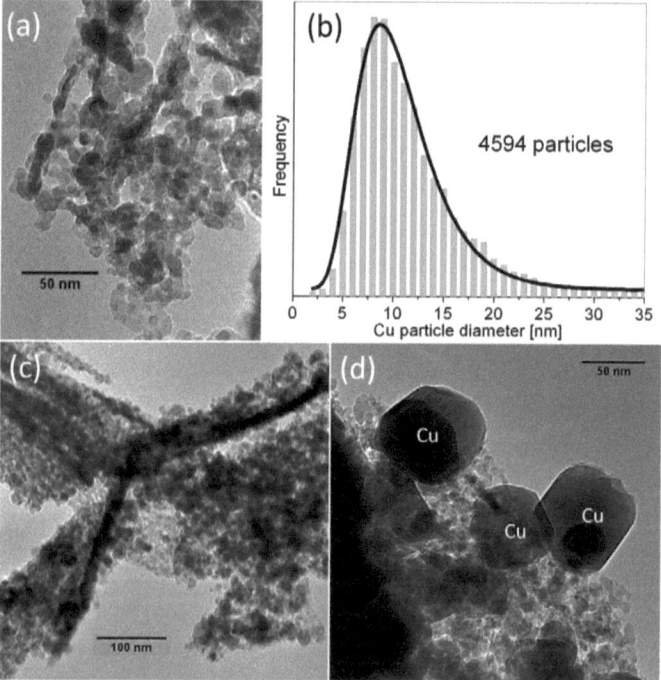

Figure 3-6: Transmission electron microscopy images of reduced Cu,Zn sample prepared at pH 6.5 (a,b) and pH 9.0 (c,d) showing the typical arrangement of Cu particles (a), the particle size distribution (b), areas before electron beam sintering (c) and fused Cu particles after electron beam sintering (d).

CZ9.0 turned out to be very unstable during the measurement. Although some images at lowest magnification were taken (Figure 3-6c), rapid partial or complete fusing of the Cu particles was observed for many areas. This effect was worse, when the electron beam was concentrated on them at higher magnification (Figure 3-6d). ZnO had a stabilizing effect against sintering because areas with high Cu:Zn ratio (76:24) were less stable than areas with low ratio (58:42). The former can be assigned to ex zincian malachite domains, the latter to ex aurichalcite. In the Zn rich stable area, the Cu particle size was 10.1 nm, but probably this value is not representative for the whole sample. Figure 3-6c shows that in agreement with the observation of Baltes et al. [11], that the Cu/ZnO aggregates still exhibit the pseudo-morphology of the zincian malachite precursor needles.

The observed differences in microstructural homogeneity are in agreement with the result of the precursor material analysis. Despite the presence of a similar amount of aurichalcite in this sample, at pH 6.5 a homogeneous microstructure is observed. The major fraction of the material stems from the zincian malachite precursor and the local compositions are similar to those determined for this precursor phase by XRD. These domains exhibit a fine distribution of Cu and ZnO particles with a unimodal Cu particle size distribution. In case of the sample prepared at pH 9, the ex-zincian malachite domains are unstable in the electron beam, which is tentatively related to the very Cu-rich zM2 phase, which cannot offer much stabilization of the Cu nanoparticles against sintering by inter-dispersion with ZnO.

Cu surface area determination of the reduced catalysts with N_2O-RFC revealed a maximum averaged value of around 25 m^2 g_{calc}^{-1} for CZ8.5 (Figure 3-7). All other pH values led to Cu surface areas in a small region between 18 and 20 m^2 g_{calc}^{-1}. CZ6.0 had the largest CuO domain size and showed the lowest Cu surface area. This can be explained with the large amount of the aurichalcite by-phase in this sample, which acts as a sink to Zn, which is in turn not available as a stabilizer against sintering in the ex-zincian malachite domains. It was surprising that CZ8.5 has a Cu surface area completely different from CZ8.0 or CZ7.0. The observed clear maximum in copper surface area for the sample CZ8.5 came as a surprise as it was not predictable from any characteristics of the precursor or the calcined material. However, because of the inhomogeneity of this sample and the complexity of possible compensating effects on the macroscopic observables of the different areas in the materials possessing a different microstructure, it will be very difficult to find the reason for the maximal copper surface area of this sample on basis of the available data.

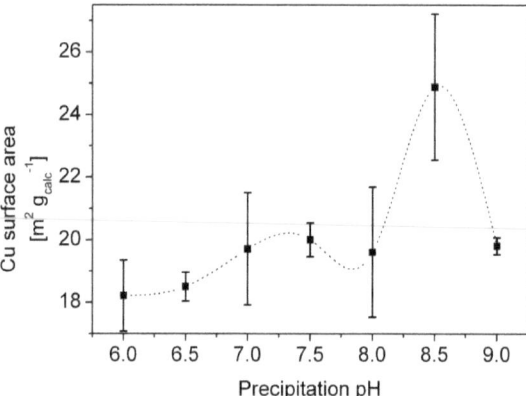

Figure 3-7: Cu surface areas of reduced Cu,Zn samples with respect to the calcined sample mass. Shown values are representing average values, endings of the error bars represent the real values obtained for the two samples for each precipitation pH value. The dotted line is a guide for the eyes.

Because the Cu surface area is known to scale linearly with the activity in methanol synthesis within the same Cu,Zn catalytic system [24], the highest activity is expected for samples prepared at pH 8.5. Accordingly, the intrinsic activities, which are a measure for the density of catalytically active sites on the surface of the Cu particles, are expected to be similar for all samples.

Literature data from ref. [10-11] revealed precipitation pH values around 7 to yield the largest Cu surface areas and activity in methanol synthesis. However, this data was obtained for Cu,Zn,Al systems which might behave different from the binary Cu,Zn system applied in this study.

3.4 Conclusions

It was shown that the variation of the pH value during the precipitation of Cu,Zn hydroxy carbonate precursors (Cu:Zn = 70:30) had an influence on the subsequent aging process and also on the ratio of the obtained precursor phases. Rietveld analysis was performed on the XRD pattern of the precursors. For pH 6.0, large fractions of the by-phase aurichalcite were found beyond zincian malachite as the main phase. Higher pH values decreased the aurichalcite fraction, with the consequence, that more Zn was introduced into the zincian malachite phase, visible from the shifted position of the $20\bar{1}$ reflection. Samples prepared at pH 7.5 and higher showed a split up signal of the $20\bar{1}$ reflection indicating two clearly different Zn substituted zincian malachite phases. Samples prepared at $6.0 \leq pH \leq 7.0$ showed a better homogeneity of the Zn distribution within the zincian malachite. As a consequence of the Zn incorporation, the calcined samples prepared at pH 6.0 showed the largest CuO domain size. Cu surface areas of the reduced Cu/ZnO catalysts revealed similar values in the region of 18 to 20 m^2 g^{-1}. Only samples prepared at pH 8.5 showed a larger Cu surface area of around 25 m^2 g^{-1}. Activity in methanol synthesis was not measured but is expected to scale linearly with the Cu surface area.

This study has shown the complexity of a catalyst synthesis by co-precipitation. The properties of the precursor materials obtained by aging of the co-precipitate have decisive influence on the structural properties and performance of the final catalysts. Unfortunately, directly tracking back the catalytic performance to the synthesis pH in a simple synthesis parameter–structure–performance relationship was found to be very complex as variation of the parameter pH induced numerous changes in the precursor material that lead to different and partially compensating effects for the resulting catalyst.

Acknowledgement

This paper has emerged from a joint research project "Next generation methanol Frank Girgsdies (help with XRD pattern analysis), Edith Kitzelmann (XRD measurements), Achim Klein-Hoffmann and Olaf Timpe (XRF) and Gisela Lorenz (BET measurements) are acknowledged. Financial support was given by the German Federal Ministry of Education and Research (BMBF, FKZ 01RI0529, 2005-2008) and the STMWFK (NW-0810-0002, since 2010). Robert Schlögl is greatly acknowledged for valuable discussions and his continuous support.

3.5 References

[1]   D. Waller, D. Stirling, F. S. Stone, M. S. Spencer, *Faraday Discuss.* **1989**, *87*, 107-120.
[2]   M. S. Spencer, *Catal. Lett.* **2000**, *66*, 255-257.
[3]   B. Bems, M. Schur, A. Dassenoy, H. Junkes, D. Herein, R. Schlögl, *Chem. Eur. J.* **2003**, *9*, 2039-2052.
[4]   A. M. Pollard, M. S. Spencer, R. G. Thomas, P. A. Williams, J. Holt, J. R. Jennings, *Appl. Catal. A* **1992**, *85*, 1-11.
[5]   E. N. Muhamad, R. Irmawati, Y. H. Tautiq-Yap, A. H. Abdullah, B. L. Kniep, F. Girgsdies, T. Ressler, *Catal. Today* **2008**, *131*, 118-124.
[6]   H. Jung, D.-R. Yang, O.-S. Joo, K.-D. Jung, *B. Kor. Chem. Soc.* **2010**, *31*, 1241-1246.
[7]   S. Kaluza, M. Behrens, N. Schiefenhoevel, B. Kniep, R. Fischer, R. Schlögl, M. Muhler, *Chem. Cat. Chem.* **2011**, *3*, 189-199.
[8]   S. Fujita, A. M. Satriyo, G. C. Shen, N. Takezawa, *Catal. Lett.* **1995**, *34*, 85-92.
[9]   M. Behrens, D. Brennecke, F. Girgsdies, S. Kißner, A. Trunschke, N. Nasrudin, S. Zakaria, N. F. Idris, S. B. Abd Hamid, B. Kniep, R. Fischer, W. Busser, M. Muhler, R. Schlögl, *Appl. Catal. A* **2011**, *392*, 93-102.
[10] J. L. Li, T. Inui, *Appl. Catal. A* **1996**, *137*, 105-117.
[11] C. Baltes, S. Vukojevic, F. Schüth, *J. Catal.* **2008**, *258*, 334-344.
[12] J. S. Campbell, *Ind. Eng. Chem. Process Des. Dev.* **1970**, *9*, 588-&.
[13] S. Zander, B. Seidlhofer, M. Behrens, *Dalton Trans.* **2012**, *41*, 13413-13422.
[14] P. Porta, S. Derossi, G. Ferraris, M. Lojacono, G. Minelli, G. Moretti, *J. Catal.* **1988**, *109*, 367-377.
[15] M. Behrens, *J. Catal.* **2009**, *267*, 24-29.
[16] A. Coelho, 4.2 ed., Bruker AXS GmbH, Karlsruhe, Germany, **2003-2009**.
[17] D. A. M. Monti, A. Baiker, *J. Catal.* **1983**, *83*, 323-335.
[18] P. Malet, A. Caballero, *J. Chem. Soc. Faraday T. 1* **1988**, *84*, 2369-2375.
[19] G. C. Chinchen, K. C. Waugh, D. A. Whan, *Appl. Catal.* **1986**, *25*, 101-107.
[20] M. Behrens, F. Girgsdies, *Z. Anorg. Allg. Chem.* **2010**, *636*, 919-927.
[21] M. Behrens, F. Girgsdies, A. Trunschke, R. Schlögl, *Eur. J. Inorg. Chem.* **2009**, 1347-1357.
[22] I. Kasatkin, P. Kurr, B. Kniep, A. Trunschke, R. Schlögl, *Angew. Chem. Int. Edit.* **2007**, *46*, 7324-7327.
[23] T. Ungar, J. Gubicza, G. Ribarik, A. Borbely, *J. Appl. Cryst.* **2001**, *34*, 298-310.
[24] M. Kurtz, H. Wilmer, T. Genger, O. Hinrichsen, M. Muhler, *Catal. Lett.* **2003**, *86*, 77-80.

Supplementary Information

Table S3-1: Denotation of samples according to their precipitation pH value. Averaged Zn content ($x_{Zn} + x_{Cu}$ = 100%) of the calcined samples from XRF. Listed values are representing average values, errors given in brackets represent the real values obtained for the two samples for each precipitation pH value.

Label	Precipitation pH value	Precursor sample number (internal)	Calcined sample number (internal)	Zn content [%]	T_{max}[a] [K]	Reducibility[b] [%]
CZ6.0	6.0	10785 10787	10786 10788	29.5 (± 0.3)	473 (± 1)	102 (± 1)
CZ6.5	6.5	7399 10084	7400 10085	29.2 (± 0.3)	477 (± 4)	99 (± 2)
CZ7.0	7.0	7192 10783	7193 10784	29.2 (± 0.2)	476 (± 2)	95 (± 3)
CZ7.5	7.5	7852 9385	7853 9386	33.8 (± 0.3)	475 (± 0)	104 (± 2)
CZ8.0	8.0	7282 7632	7283 7633	29.3 (± 0.3)	479 (± 3)	97 (± 1)
CZ8.5	8.5	7938 10147	7939 10148	29.7 (± 0.2)	475 (± 0)	103 (± 1)
CZ9.0	9.0	6501 10086	6521 10087	27.5 (± 2.1)	475 (± 1)	97 (± 1)

[a] Temperature of maximum H_2 consumption rate during TPR
[b] Hydrogen consumption relative to the CuO content after TPR

Scanning electron microscopy

SEM investigation was done on one precursor sample in each case of CZ7.0, CZ7.5 and CZ9.0 with respect to the question whether evidence for the mentioned inhomogeneity was found. Exemplarily, images of CZ9.0 showing round shaped large aggregates are depicted in Figure S3-a,b. This droplet-like shape is a result of the spray-drying of the washed precursor slurry. No obvious differences in the morphology were found within the three investigated samples.

Indeed, no homogeneous morphology of the primary particles was observed, but needles, platelets, particles and some areas with smooth surfaces were found in the samples (Figure S3-c,d). The needle like morphology was predominant and was already observed in the Cu,Zn,Al-system where this was assigned to zincian malachite [1]. All samples additionally contain different phases of zincian malachite and aurichalcite. The range of observed Cu:Zn ratios measured by EDX at different spots (at least 10 per sample) was similar for CZ7.0 and CZ7.5 (≈ from 55:45 to 75:25) and somewhat narrower for CZ9.0 (≈ from 63:37 to 72:28). Zn richer spots should be due to the presence of aurichalcite and Cu rich spots due to zincian malachite. Because in CZ7.5 the spot with the lowest Zn content measured by EDX was

25 mol%, no spot of pure zM2 with an expected Zn content of 15% (Zn poor zM2, according to XRD, Figure 3-5e) was detected. This is due to the fact that EDX cannot be applied to infinitely small regions. The elemental composition of the measured spots is probably a superposition of many small crystallites from different phases (aurichalcite, zM1, zM2) and gives rather an averaged picture than a detailed result.

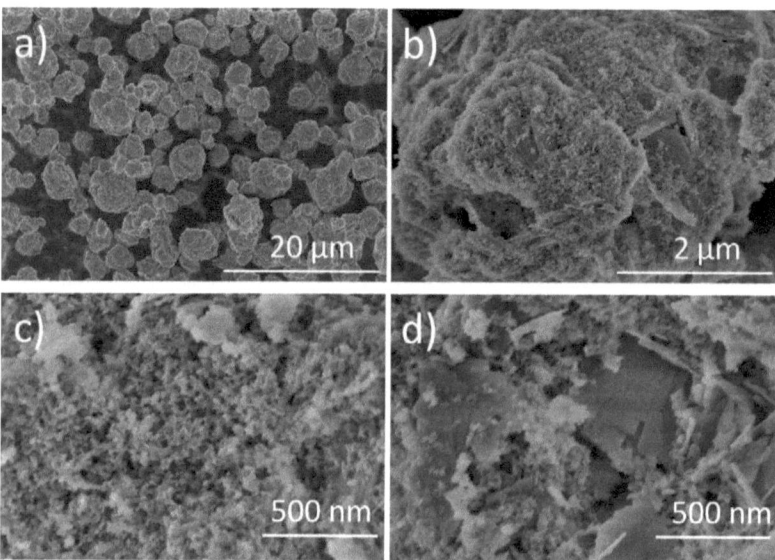

Figure S3-1: Scanning electron microscopy images (2.5 keV) of a Cu,Zn hydroxy carbonate precursor sample prepared at pH 9.0 showing an overview image with roundish primary particles (a), a primary particle (b), zincian malachite needles, platelets and particles (c), plane areas (d).

Thermal analysis

Thermal decomposition of the precursors was performed to obtain nano-sized CuO and ZnO particles. This calcination step was monitored by thermogravimetric measurements (TG) combined with evolved gas analysis (EGA). The TG-EGA experiments were executed up to 973 K, in contrast to the calcinations for catalyst preparation, which is performed up to 603 K. Anions in the Cu,Zn hydroxy carbonates are decomposed under emission of water and carbon dioxide. XRD patterns of all samples after thermal treatment up to 973 K revealed well-crystalline CuO and ZnO (not shown).

Exemplarily, the evolutions of the mass loss and the normalized H_2O and CO_2 traces for a Cu,Zn hydroxy carbonate precursor sample prepared at pH 7.5 are depicted in Figure S3-2. EGA shows that the decomposition mainly proceeds in three steps. After the release of physisorbed or incorporated water (range I, up to 403 K), the second step is characterized by simultaneous emission of H_2O and CO_2 (range II, ca. 403-673 K). In the third step, only CO_2 is emitted at high temperatures (range III, ca. 673-873 K). The reason of this last decomposition step is the presence of temperature stable carbonate species (HT-CO_3^{2-}) which are probably located at the interface between the formed CuO and ZnO [2-3]. The role of this residual carbonate, which is still present after calcination at not too high temperatures, is debated and it has been proposed that it can stabilize oxidized copper species in the reduced catalyst and increase the activity [4]. The abundance and the thermal stability of these species are suggested to be a measure for the amount and the strength of the interactions across interfaces and grain boundaries of CuO/ZnO aggregates. Therefore, pure malachite $Cu_2(OH)_2CO_3$ and hydrozincite $Zn_5(OH)_6(CO_3)_2$ do not show the emission of these species [2-3, 5].

The theoretical mass losses for zincian malachite $(Cu_{1-x}Zn_x)_2(OH)_2CO_3$ and aurichalcite $(Cu_{1-y}Zn_y)_5(OH)_6(CO_3)_2$ can be calculated to 28% and 26%, respectively, and depend only slightly on the Cu:Zn ratio because of the similar molar masses of Cu and Zn. Thus, samples containing low zincian malachite fractions according to XRD should show low mass losses. Indeed, CZ6.0 exhibits the lowest value of 26.9 wt% (Figure S3-3a) and all other precipitation pH values show higher mass losses (28.2 to 29.2 wt%) due to lower aurichalcite fractions.

The HT-CO_3^{2-} decomposition temperatures indicate the strength of interaction between CuO and ZnO particles. The averaged values are in the range from around 720 (CZ7.5 and CZ8.0) to 750 K (CZ8.5) but no simple trend can be observed (Figure S3-3b) and the error bars are relatively large. The HT-CO_3^{2-} amount can be calculated from the fraction of HT-CO_2 related to the overall CO_2 emission in a semi-quantitative manner based on the integrals of the MS traces. This fraction is relatively constant around 48-50% for precipitation pH values up to 7.0 and ranges between 38 and 45% for higher pH values (Figure S3-3c). The smallest proportions are detected for pH 7.5 and 8.0 with about 40%. Theses samples also showed the lowest temperature for HT-CO_2 emission. Again, no clear trend with preparation pH can be observed. Because the calcination of the precursors is performed only at 603 K, the high temperature carbonate species stays present in the sample. It was suggested that it can contribute to increased activity by the formation of copper suboxide species [4].

Chapter 3: Supplementary Information

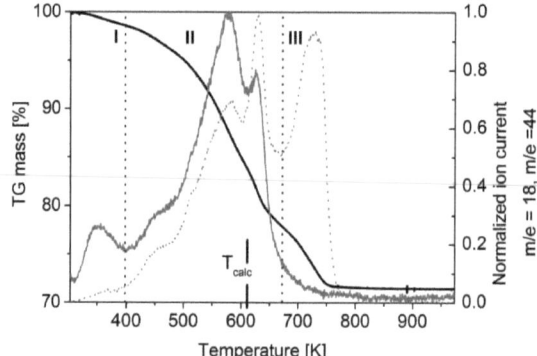

Figure S3-2: TG-MS results of hydroxy carbonate precursor sample prepared at pH 7.5: mass loss (black), MS traces of H_2O (grey, solid line) and CO_2 (grey, dotted line).

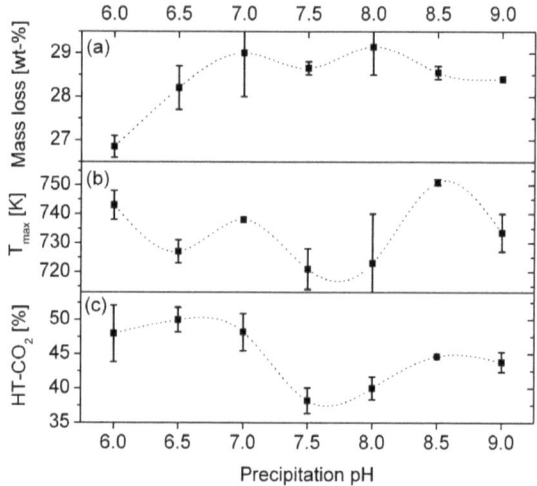

Figure S3-3: TG-MS results of Cu,Zn hydroxy carbonate precursors: mass loss after heating to 973 K (a), temperature of highest CO_2 emission rate (b) and CO_2 emission above 673 K relative to overall CO_2 emission according to MS trace (c). Shown values are representing average values, endings of the error bars represent the real values obtained for the two samples for each precipitation pH value. Lines are guides for the eyes.

Temperature programmed reduction

Description of the reduction profiles with a single peak was not possible because at least one shoulder was observed (Figure S3-). This can be attributed to the reduction of CuO in multiple steps [6], reduction of multiple CuO species, e.g. from different precursor phases [7], or reduction of other components than CuO.

All reduction profiles were analyzed with respect to the temperature of the highest H_2 consumption (473-479 K) and the reducibility of CuO (95-104%). The results were summarized in Table S3-1. Within the series, neither distinct changes nor clear trends were observed.

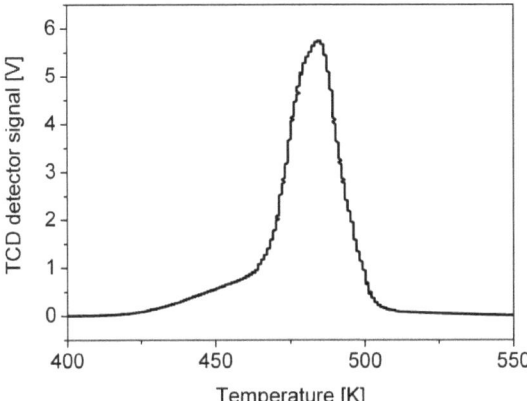

Figure S3-4: TPR profile of a calcined Cu,Zn sample prepared at pH 6.5.

[1] M. Behrens, *J. Catal.* **2009**, *267*, 24-29.
[2] B. Bems, M. Schur, A. Dassenoy, H. Junkes, D. Herein, R. Schlögl, *Chem. Eur. J.* **2003**, *9*, 2039-2052.
[3] G. J. Millar, I. H. Holm, P. J. R. Uwins, J. Drennan, *J. Chem. Soc. Faraday Trans.* **1998**, *94*, 593-600.
[4] L. M. Plyasova, T. M. Yureva, T. A. Kriger, O. V. Makarova, V. I. Zaikovskii, L. P. Soloveva, A. N. Shmakov, *Kinet. Catal.* **1995**, *36*, 425-433.
[5] M. Behrens, F. Girgsdies, A. Trunschke, R. Schlögl, *Eur. J. Inorg. Chem.* **2009**, 1347-1357.
[6] M. M. Günter, B. Bems, R. Schlögl, T. Ressler, *J. Synchrotron Rad.* **2001**, *8*, 619-621.
[7] G. Fierro, M. LoJacono, M. Inversi, P. Porta, F. Cioci, R. Lavecchia, *Appl. Catal. A* **1996**, *137*, 327-348.

Chapter 4: The Role of the Oxide Component in the Development of Copper Composite Catalysts for Methanol Synthesis

Stefan Zander, Edward Kunkes, Manfred E. Schuster, Julia Schumann, Gisela Weinberg, Robert Schlögl, Malte Behrens.

Abstract

In this work we compare the classical zincian malachite-derived Cu/ZnO with a new Cu/MgO catalyst at a fixed molar ratio of Cu to Zn and Mg, respectively, of 80:20. The geometric influence of MgO turned out to be better compared to ZnO but the synergetic effect of Cu and ZnO during methanol synthesis from $CO_2/CO/H_2$ was unequaled. Both geometric and synergetic effects were combined by preparation of Cu/MgO/ZnO sample which exhibited a higher activity than Cu/ZnO and Cu/MgO. Changing the feed gas to CO/H_2, Cu/MgO was most active.

4.1 Introduction

The development and optimization of industrially applied high performance catalysts is usually a continuous process that to a large extent is based on the experience of the manufacturer. The accumulated knowledge from the combination of empirical trial-and-error experimentation and an afterward structure-function-relationship-guided optimization approach within the boundary conditions of a feasible and scalable synthesis often leads to very complex recipes – sometimes generalized as the "black magic" of catalyst preparation. In the last years, we have worked on better understanding the synthesis and functionality of the $Cu/ZnO/Al_2O_3$ methanol synthesis catalyst using the well-documented industrially applied preparation route[1-4] as starting point. As a result of this effort, we have elaborated a model of the so-called "chemical memory"[5-6] of catalyst preparation and of the active site in this catalyst system.[7] Here, we show how this knowledge can be used to develop a new family of Cu-based catalysts.

With the help of structure-performance-relationships observed within a series of functional powder catalysts and DFT calculations, the active site of industrial methanol synthesis was identified as a combination of a surface defect of Cu and the presence of partially reduced Zn species at this defect,[7] explaining the widely studied "Cu-ZnO synergy".[8-9] Within the industrial synthesis, high concentrations of these sites can be realized by preparation of defective Cu nanoparticles and migration of ZnO_x species onto the Cu surface as a result of strong-metal-support interaction (SMSI)[7, 10-12] and an intimate interface contact of both catalyst components. At the same time, the total accessible Cu surface area (SA_{Cu}) is large, because the bulk-catalyst is prepared with a porous microstructure[5, 13] from a co-precipitated precursor compound. In this context, ZnO acts as a geometrical spacer between the Cu nanoparticles and helps to increase and stabilize the Cu dispersion.[9, 14] Thus, ZnO has two functionalities in the final catalyst: (i) as nanoparticles it acts as a physical spacer between the Cu particles stabilizing the porous microstructure and (ii) as a thin layer at the surface of the Cu particles it is an essential ingredient for the active site. The work presented here was guided by the idea to separate these two effects.

A simplified scheme of the relevant properties of Cu/ZnO methanol synthesis catalysts is shown in Figure 4-1a. Three prerequisites have to be fulfilled in order to generate a high performing catalyst. The material should have a high SA_{Cu} to expose a large number of active sites; the Cu phase must be defective to achieve a high density of active sites at the surface; and SMSI-induced synergetic effect of ZnO must be present to activate the defect sites for methanol synthesis. Only if all three factors come together, in the darkest shaded region of Figure 4-1a, the catalyst will be highly active for methanol synthesis. The defects are generated by the careful and delicate preparation method yielding distorted Cu nanoparticles in close contact to the oxide phase, while the other two properties are governed by function (i) and (ii) of the ZnO component.

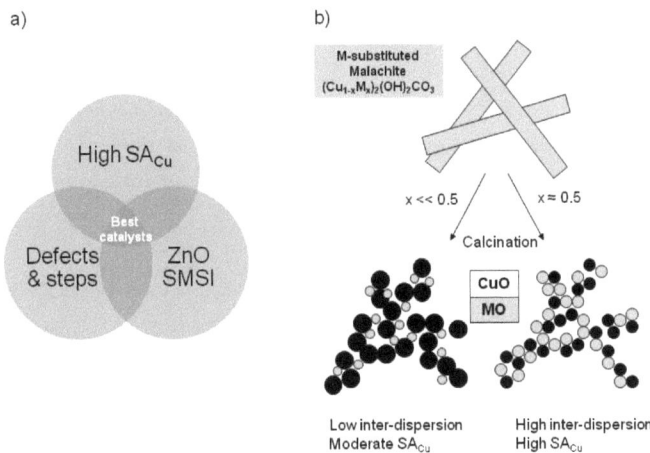

Figure 4-1: a) Schematic representations of the necessary ingredients for a high performance methanol synthesis catalyst. b) Scheme of the role of precursor composition for the Cu dispersion in the final catalyst.

The synthesis route for preparing Cu/ZnO/(Al$_2$O$_3$) catalysts follows a multi-step procedure including T- and pH-controlled co-precipitation[6] of aqueous Cu,Zn,Al nitrate solution with sodium carbonate solution followed by aging,[15] washing, and drying to yield a hydroxide-carbonate precursor. This material is calcined and finally activated by reduction of CuO to metallic Cu. Low amounts of Al$_2$O$_3$ acts as a structural promoter in the industrially applied catalyst.[16-17]

The relevant precursor material has been identified as thin needles of zincian malachite, (Cu,Zn)$_2$(OH)$_2$CO$_3$.[5] The incorporation of Zn^{2+} into the cationic lattice of malachite favors the nano-structuring of the CuO/ZnO aggregates formed upon calcination due to the perfect distribution of both species in the joint crystal lattice of the precursor compound. This can be understood as a purely geometric effect, which is schematically shown in Figure 4-1b and is the basis for the functionality (i) of ZnO. Zn^{2+} is well suited for this purpose, because it exhibits the same charge and an ionic radius similar to Cu^{2+} favoring substitution. However, incorporation into the malachite lattice is limited to < 30% due to solid state chemical constraints[18] that are likely due to the differences in the coordination environment between the Jahn-Teller-ion Cu^{2+} (d^9) and Zn^{2+} (d^{10}). Mg^{2+} is an interesting replacement for Zn^{2+}, because its charge matches and its ionic radius differs, alike Zn^{2+}, by less than 2% from that of Cu^{2+}. Furthermore, $(Cu,Mg)_2(OH)_2CO_3$ crystallizes in the rosasite crystal structure, which is closely related to that of malachite and should open the door for a comparable precursor chemistry between Cu,Zn and Cu,Mg compounds. Moreover, $(Cu,Mg)_2(OH)_2CO_3$ is naturally occurring as the mineral McGuinessite[19-20] and can incorporate more Mg^{2+} than Cu^{2+}, which has not been achieved for zincian malachite. Thus, a more efficient dilution of the Cu^{2+} ions might be possible with Mg^{2+} compared to Zn^{2+} by application of lower amounts of Cu to further promote the nanostructuring and increase the Cu dispersion.

In this work we compare the classical zincian malachite-derived Cu/ZnO with a new Cu/MgO catalyst at a fixed molar ratio of Cu to Zn and Mg, respectively, of 80:20. At this ratio, Zn incorporation into malachite does not exceed the critical Zn concentration in zincian malachite to assure synthesis of phase-pure precursor compounds resulting in high comparability of the Cu,Zn and Cu,Mg systems and in uniform catalysts whose properties can be easily traced back to the precursor compounds. Both precursors were prepared from mixed nitrate solutions by controlled co-precipitation with sodium carbonate solution and subsequent ageing in the mother liquor. They are denoted CZ and CM in the following.

4.2 Experimental

4.2.1 Catalyst Preparation

Hydroxide-carbonate precursors of CZ and CM were synthesized by co-precipitation (T = 338 K) from Cu,Zn and Cu,Mg (80:20) nitrate solutions and Na_2CO_3 solution as precipitating agent in an automated lab reactor (LabMax, Mettler Toledo). The pH was set to 6.5 for CZ and 9.0 for CM. The precipitates were aged (> 60 min), filtrated, washed and dried. Calcination was carried out in air at 603 K (2 K min^{-1}) for 3 h. One part of the calcined CM was impregnated with Zn citrate solution, dried and calcined again at the same T.

4.2.2 Characterization

X-ray fluorescence (XRF) was performed using a Bruker S4 Pioneer X-ray spectrometer. XRD was measured on a STOE STADI P transmission diffractometer using Cu K$_{\alpha 1}$ radiation. Specific surface areas were determined by N_2 physisorption at liquid nitrogen T in a Quantachrome Autosorb-6 machine. SA$_{Cu}$ was determined in a custom-made setup by applying N_2O reactive frontal chromatography (N_2O-RFC) at 303 K[21-22] after reducing the catalysts at 523 K in 5% H_2. For TEM investigations, a FEI Titan Cs 80-300 microscope operated at 300 kV and energy-dispersive X-ray (EDX) analyzer was used. Spherical aberrations were corrected by use of the CEOS Cs-corrector reaching an information limit of 0.8 Å. HRTEM pictures were processed to obtain the power spectra which were used to measure interplanar distances and angles for phase identification. The reduced samples were transferred in a glovebox to the microscope without contact to air.

4.2.3 Catalytic performance

Catalytic tests in CO_2 hydrogenation were carried out in a fixed bed flow reactor. 200 mg (100-200 µm, mixed with 2 g of crushed SiO_2 chips) were loaded into a 10 mm inner diameter stainless steel reactor tube. The catalysts were reduced at 573 K (1 K min^{-1}) for 1.5 hours in 20% H_2 in He. Upon completion of the reduction, the reactor was cooled to 523 K, a 3:1 H_2/CO_2 mixture (100 mL min^{-1}) containing 4% Ar (as internal standard) was introduced into the reactor, and the pressure was raised to 30 bars. Online analysis of products was performed with a gas chromatograph (Agilent 7890A). After the start of the reaction, the catalysts were allowed to stabilize for 6 hours time on stream at 523 K. CO hydrogenation was performed at the same T and p in the same set up, but using a 6 mm stainless steel reactor tube and 50 mg of catalyst diluted with 0.7 g SiO_2. The feed gas contained 14% CO, 59% H_2, 4% Ar and balance He. Performance under synthesis gas conditions was determined after switching the gas composition gradually to a composition of 6% CO, 8% CO_2, 59% H_2 and balance He.

4.3 Results and discussion

XRD of the precursors confirmed the formation of single phase materials with a crystal structure similar to malachite (Figure 4-2a). In comparison to the literature pattern of malachite, a pronounced shift of the 20-1 peak is seen in both compounds. This is an indication for the incorporation of non-Jahn-Teller cations in the lattice of malachite and from the similar position of the reflections in both patterns a similar degree of substitution can be estimated (Table 4-1) suggesting that the non-Jahn-Teller ions have been completely incorporated into the malachite structure in both samples. It is noted that CM exhibits significantly broader XRD peaks indicative of small crystallites. Also the particle morphology of CM looks rather spongy compared to CZ, which exhibits larger and well-separated particles (Figure 4-3). Accordingly, a larger BET surface area of the CM precursor has been observed (Table 4-1).

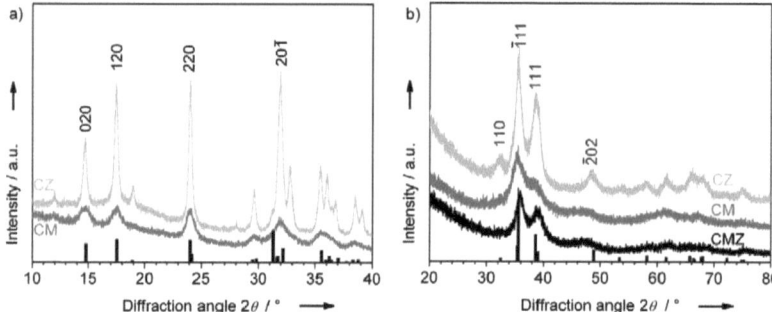

Figure 4-2: a) XRD patterns of the precursor materials of CZ (light grey) and CM (grey). The pattern of malachite (black; ICSD: 72-75) is shown as bar graph. b) XRD patterns of the calcined samples CZ (light grey), CM (grey) and CMZ (black). CuO (black; ICSD: 80-76) is included as reference.

Figure 4-3: Scanning electron microscopy images (2.5 keV) of CZ (a) and CM (b)

Table 4-1: Properties of the CZ, CM and CMZ catalysts (prec. = precursor material, calc. = calcined material).

Sample	Cu:M ratio	$D^{[c]}$ / nm prec. // calc.	SA_{BET} / m^2g^{-1} prec. // calc.	$SA_{Cu}^{[d]}$ / $m^2g_{cat}^{-1}$
CZ	80:20[a]	26.6 // 5.8	36 // 83	16.0
CM	83:17[a]	8.0 // 2.8	81 // 73	20.3
CMZ	80:16:4[b]	8.0 // 3.9	81 // 80	24.2

[a] molar, determined by XRF, ± 1 mol%
[b] molar, estimated
[c] Crystallite domain size of $(Cu,M)_2(OH)_2CO_3$ (precursor) and CuO/MO (calcined), ± 0.2 nm derived from XRD peak profiles
[d] Specific Cu surface area of the reduced catalyst determined by N_2O-chemisorption, ± 1 m^2g^{-1}

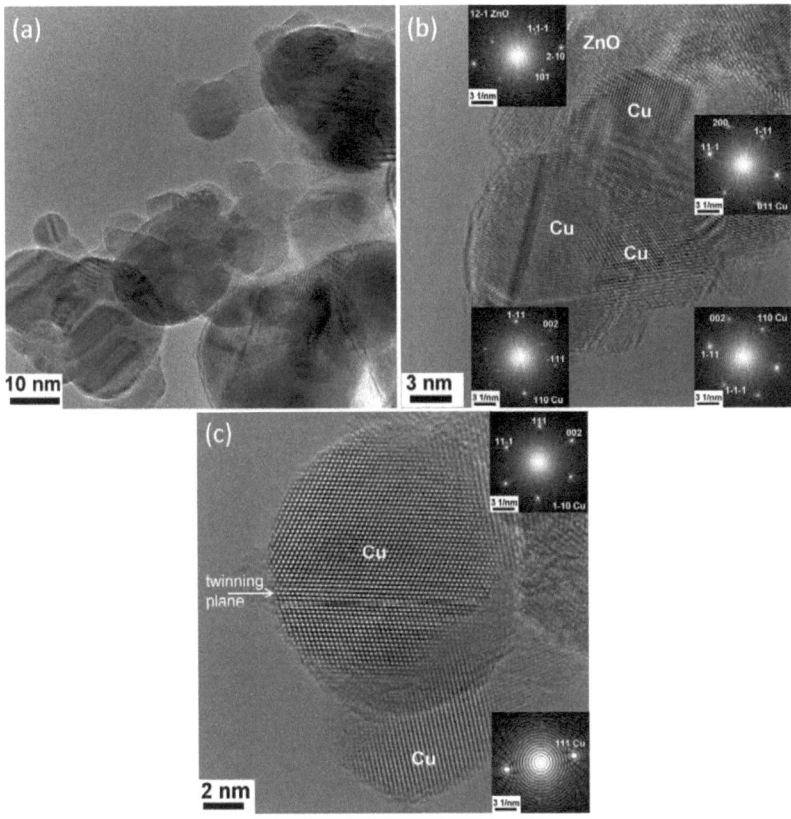

Figure 4-4: (HR-)TEM images of the reduced CZ catalyst. The insets show power spectra of the neighbouring particles and are used for phase identification.

Upon calcination at 603 K, poorly crystalline CuO is formed in both samples as evidenced by XRD (Figure 4-2b), while the ZnO and MgO components are mostly X-ray amorphous. Again

Chapter 4: Development of Cu-Catalysts for Methanol Synthesis from CO_2 and CO

CM exhibits a significantly smaller crystallite size according to the XRD peak width, but a slightly smaller specific surface area (Table 4-1). Furthermore, CM yields a by more than 30% higher specific Cu surface area after reduction.

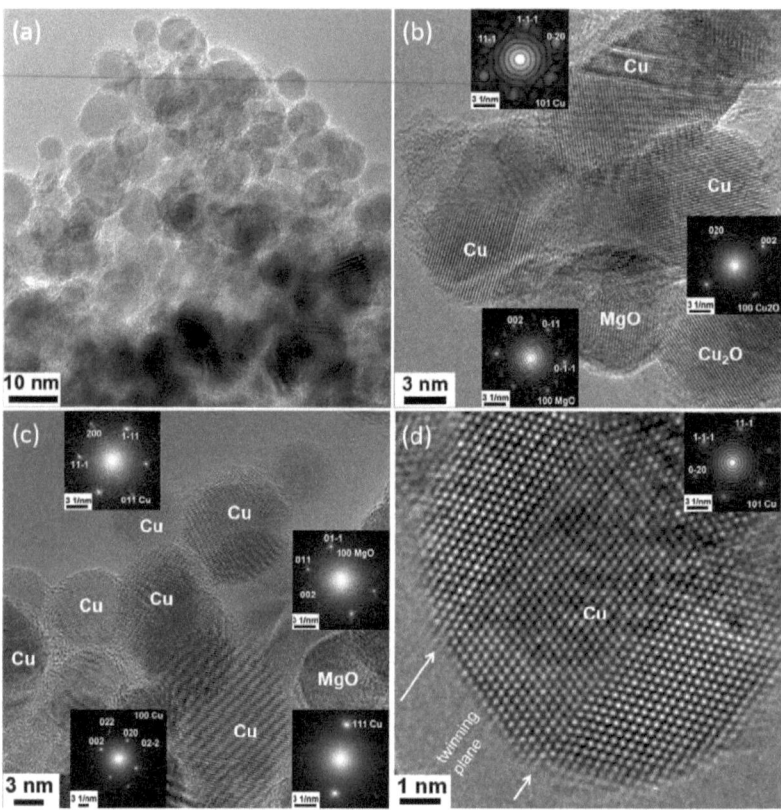

Figure 4-5: (HR-)TEM images of the reduced CM catalyst.

TEM investigation of the reduced catalysts showed that the general microstructure of CZ (Figure 4-4a,b) and CM (Figure 4-5a,b,c) is similar and characterized by arrangements of round shaped Cu particles separated by differently sized crystallites of ZnO or MgO, respectively. A few Cu_2O particles are found that probably have formed due to slight re-oxidation of the reduced catalysts upon sample transfer. The presence of larger Cu particles in CZ compared to CM is consistent with the difference in Cu surface areas (Table 4-1). This result indicates that

Chapter 4: Development of Cu-Catalysts for Methanol Synthesis from CO_2 and CO

MgO is intrinsically a better geometrical spacer compared to ZnO as even at the non-ideal substitution level of 20% Cu particles can be obtained that with an average size below 10 nm are similarly small as found in state-of-the-art $Cu/ZnO/Al_2O_3$ catalysts. Thus, the structurally promoting role of ZnO has been successfully replaced with MgO.

High resolution TEM showed that the Cu particles in both samples contain planar defects, which have been shown to contribute to the methanol synthesis activity in Cu/ZnO catalysts[7] (CZ: Figure 4-4c; CM: Figure 4-5d). Thus, the important defectiveness of Cu is probably a result of the precursor decomposition approach common to both catalysts, which leads to crystallization of distorted Cu crystallites at relatively mild T.

Both catalysts have been tested in methanol synthesis with various feed gas composition, i.e. hydrogenation of pure CO_2, a CO_2/CO mixture and pure CO. The results are shown in Figure 4-6. In the hydrogenation of pure CO_2, CZ showed a much higher activity than CM showing clearly that the methanol synthesis rate is not only a function of the exposed Cu surface area alone (Figure 4-6a). According to Figure 4-1, the low activity of CM can be explained with the absence of the synergetic SMSI-effect as MgO is an irreducible oxide that does not show SMSI in the relevant T regime. The situation is similar if methanol is produced from a typical synthesis gas mixture with CO_2 and CO in the feed (Figure 4-6b). CZ shows a slightly lower rate of methanol production compared to the CO_2/H_2 feed, while CM remains essentially inactive despite the large exposed Cu surface area. These results strikingly confirm the crucial synergetic role of the ZnO-promoter that has been subject of many literature reports.

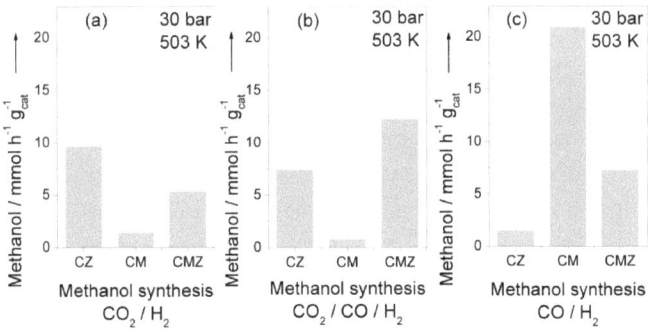

Figure 4-6: Catalytic results of methanol synthesis of the CZ, CM and CMZ catalysts in different feed gas compositions at 30 bar and 503 K.

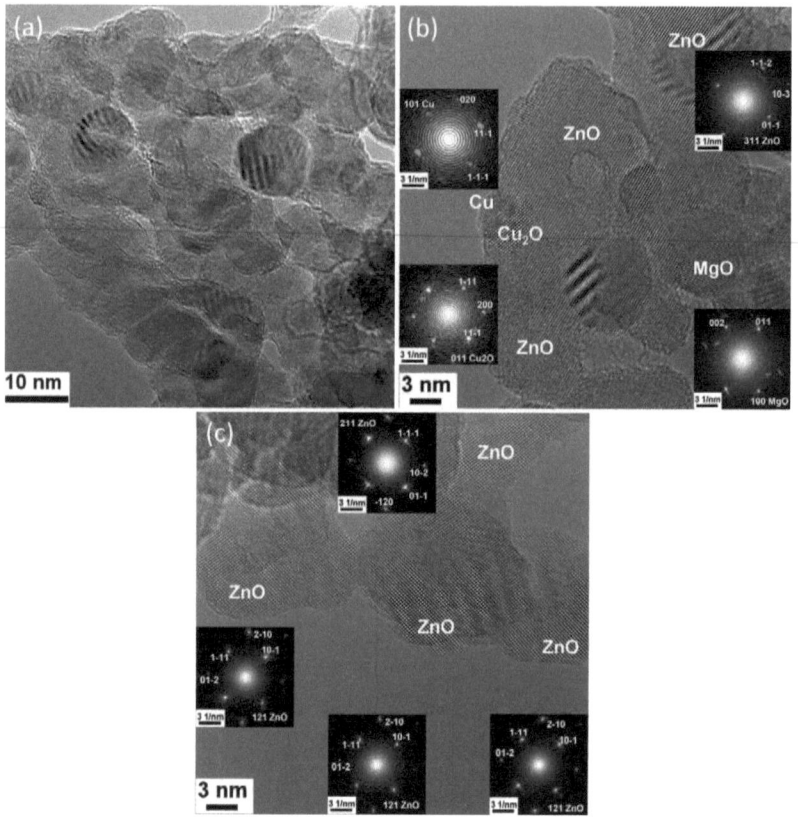

Figure 4-7: (HR-)TEM images of the reduced CMZ catalyst.

With the idea to "switch on" the lacking Cu-ZnO synergy by addition of Zn – a similar has been reported earlier for model catalysts [17] – a third catalyst, labeled CMZ, was prepared. Impregnation of the calcined CM with 5 wt% ZnO resulted in a catalyst that was indeed able to convert CO_2 and the synthesis gas mixture into methanol at much higher rates than CM (Figure 4-6a,b). HRTEM investigation of CMZ (Figure 4-7) showed similar general microstructure and particle morphology like CM (Figure 4-5) but additionally confirmed the presence of ZnO at the surface of the Cu/MgO aggregates (Figure 4-7c). The weight-based methanol production rate from synthesis gas of CMZ was even higher than that of CZ, which is a result of the better Cu dispersion in the Cu/MgO-aggregates. The intrinsic rates per SA_{Cu} are very similar for CZ and CMZ in this experiment (not shown), which points to another advantage of using MgO for Cu dispersion instead of ZnO, namely its lower specific weight. At the same molar Cu content of

80% (metal-based) the effective Cu loading of CM is as high as 86 wt%, while that of CZ is 75 wt%. The very high value for CM shows that the microstructure of this kind of catalyst should not be described as a classical supported system, but its sponge-like microstructure also resembles an "oxide-stabilized Raney-Cu".

These results show that the functions of the oxide component can be successfully separated in Cu-based methanol synthesis catalysts. It was shown for a given catalyst composition as a proof-of-principle that this approach enables preparation of high-performance catalysts and leaves additional degrees of freedom for future optimization. In particular, Cu dispersion can be optimized within the proven malachite-precursor method by increasing the Cu substitution without being bound to the constraints of the Cu,Zn system. Furthermore, the method of addition and amount of the synergistic promoter can be varied for a given highly-dispersed Cu/oxide system to switch on the production of methanol from CO_2 or synthesis gas. In CMZ areas of relatively large particles of very crystalline ZnO have been observed (Fig. 6c). These domains probably do not contribute to the synergetic catalytic effect and point to the possibility to further improve this catalyst by optimization of the ZnO addition.

Interestingly, the catalytic performance of the samples is completely changed when a CO/H_2 feed is used for methanol synthesis. Here CM shows a very high methanol production rate, which clearly exceeds that of CZ or CMZ in the other feed gases (Figure 4-6c). This result is in line with previous studies [23-24] that have shown that MgO-supported Cu is a powerful CO hydrogenation catalyst. Interestingly, while it was a prerequisite for methanol production in CO_2-containing feeds, the addition of Zn to CM was detrimental in this reaction possibly by partially covering of the active surface. Thus, in addition of being a very powerful CO hydrogenation catalyst, the Cu/MgO and derived Cu/MgO/ZnO systems also represent a suitable material basis for conducting basic studies on the roles of synergy, dispersion and structural dynamics for methanol synthesis in different feed gases.

In summary, the high comparability of the three catalysts due to the similar general morphology found by TEM investigation, allows tracing back the differences in activity of the samples to the influence of the oxide phase(s) ZnO and/or MgO. These two oxides do not only act as structural promoters, but also determine the preferred pathway of methanol synthesis from CO_2 or CO as carbon source.

4.4 Conclusion

We propose that the presented synthetic approach opens the door to exploit new room for knowledge-based optimization of the proven Cu/ZnO/Al$_2$O$_3$ catalyst system. It seems in particular promising to combine the highest possible substitution of Cu^{2+} in malachite by a suitable diluent like Mg^{2+} for proper structural promotion with the right amount of a reducible oxide at the surface of the Cu particle like ZnO for proper synergistic promotion. Furthermore, the presented materials show potential to fertilize new progress in the long-lasting studies of the mechanism of methanol synthesis by providing fundamental insight into the role of different material components. Future studies in our lab will aim at better understanding the observed differences, which point to fundamental differences in the active sites for CO and CO$_2$ conversion over the same catalysts, in particular with regard to the involvement of Zn in the latter, but not in the former. Thus, in addition of being a very powerful CO hydrogenation catalyst, the Cu/MgO and derived Cu/MgO/ZnO systems also represent a suitable material basis for conducting basic studies on the roles of synergy, dispersion and structural dynamics for methanol synthesis in different feed gases.

Acknowledgement

Edith Kitzelmann (XRD measurements), Achim Klein-Hoffmann and Olaf Timpe (XRF), Gisela Lorenz (BET measurements), Nygil Thomas (help with catalytic measurements) are acknowledged. Financial support was given by the German Federal Ministry of Education and Research (BMBF, FKZ 01RI0529, 2005-2008) and the STMWFK (NW-0810-0002, since 2010).

4.5 References

[1] C. Baltes, S. Vukojevic, F. Schüth, *J. Catal.* **2008**, *258*, 334-344.
[2] D. Waller, D. Stirling, F. S. Stone, M. S. Spencer, *Faraday Discuss.* **1989**, *87*, 107-120.
[3] J. L. Li, T. Inui, *Appl. Catal. A* **1996**, *137*, 105-117.
[4] J. B. Hansen, P. E. H. Nielsen, in *Handbook of Heterogeneous Catalysis, Vol. 6* (Eds.: G. Ertl, G. Knözinger, F. Schüth, J. Weitkamp), Wiley-VCH, Weinheim 2nd ed., **2008**, pp. 2920-2949.
[5] M. Behrens, *J. Catal.* **2009**, *267*, 24-29.
[6] M. Behrens, D. Brennecke, F. Girgsdies, S. Kißner, A. Trunschke, N. Nasrudin, S. Zakaria, N. F. Idris, S. B. Abd Hamid, B. Kniep, R. Fischer, W. Busser, M. Muhler, R. Schlögl, *Appl. Catal. A* **2011**, *392*, 93-102.
[7] M. Behrens, F. Studt, I. Kasatkin, S. Kühl, M. Hävecker, F. Abild-Pedersen, S. Zander, F. Girgsdies, P. Kurr, B.-L. Kniep, M. Tovar, R. W. Fischer, J. K. Nørskov, R. Schlögl, *Science* **2012**, *336*, 893-897.
[8] Y. Kanai, T. Watanabe, T. Fujitani, T. Uchijima, J. Nakamura, *Catal. Lett.* **1996**, *38*, 157-163.
[9] M. S. Spencer, *Top. Catal.* **1999**, *8*, 259-266.
[10] Y. Kanai, T. Watanabe, T. Fujitani, M. Saito, J. Nakamura, T. Uchijima, *Catal. Lett.* **1994**, *27*, 67-78.
[11] J. D. Grunwaldt, A. M. Molenbroek, N. Y. Topsoe, H. Topsoe, B. S. Clausen, *J. Catal.* **2000**, *194*, 452-460.
[12] R. N. d'Alnoncourt, X. Xia, J. Strunk, E. Löffler, O. Hinrichsen, M. Muhler, *Phys. Chem. Chem. Phys.* **2006**, *8*, 1525-1538.
[13] I. Kasatkin, P. Kurr, B. Kniep, A. Trunschke, R. Schlögl, *Angew. Chem. Int. Edit.* **2007**, *46*, 7324-7327.
[14] T. Fujitani, J. Nakamura, *Catal. Lett.* **1998**, *56*, 119-124.
[15] S. Zander, B. Seidlhofer, M. Behrens, *Dalton Trans.* **2012**, *41*, 13413-13422.
[16] M. V. Twigg, M. S. Spencer, *Top. Catal.* **2003**, *22*, 191-203.
[17] M. Kurtz, N. Bauer, C. Buscher, H. Wilmer, O. Hinrichsen, R. Becker, S. Rabe, K. Merz, M. Driess, R. A. Fischer, M. Muhler, *Catal. Lett.* **2004**, *92*, 49-52.
[18] M. Behrens, F. Girgsdies, *Z. Anorg. Allg. Chem.* **2010**, *636*, 919-927.
[19] N. Perchiazzi, *Z. Kristallogr.* **2006**, 505-510.
[20] M. Fleischer, L. J. Cabri, *Am. Mineral.* **1981**, *66*, 1274-1280.
[21] G. C. Chinchen, C. M. Hay, H. D. Vandervell, K. C. Waugh, *J. Catal.* **1987**, *103*, 79-86.
[22] O. Hinrichsen, T. Genger, M. Muhler, *Chem. Eng. Technol.* **2000**, *23*, 956-959.
[23] B. Denise, R. P. A. Sneeden, *Appl. Catal.* **1986**, *28*, 235-239.
[24] J. C. J. Bart, R. P. A. Sneeden, *Catal. Today* **1987**, *2*, 1-124.

Supplementary Information

Table S4-1: Internal sample numbers

Sample	Precursor	Calcined	Reduced
CZ	7749	7750	13284
CM	9278	10639	13121
CMZ	(9278)	13188	13572

Chapter 5: Promoting Methanol Synthesis Catalysts: Correlations between Microstructure and Activity in Cu/ZnO/Ga$_2$O$_3$

Stefan Zander, Julia Schumann, Igor Kasatkin, Gisela Weinberg, Gregor Koch, Thorsten Ressler, Patrick Kurr, Benjamin Kniep, Malte Behrens.

Abstract

We report on introduction of gallia as promoter in the Cu,Zn system. Samples with different promoter content were prepared by co-precipitation. Characterization results (XRF, XRD, N$_2$ physisorption, TGMS, TPR, SEM, TEM, UV-Vis, and XANES) during different stadia of the catalyst preparation process as well as catalytic results in methanol synthesis are presented. The promoting effect is most effective for low amounts of Ga^{3+} ($\leq$ 3 mol%). An increase in absolute methanol synthesis activity of 60% compared to the binary Ga free system was observed. Promotion is characterized by a geometric modification which is expressed by a higher Cu surface area. In contrast, addition of Ga leads to slightly lower intrinsic activities (related to Cu surface area), probably by modification of the ZnO phase by incorporated Ga species and consequences for Cu-ZnO synergy. The extent of the geometric and the synergetic effect depends on incorporation of Zn^{2+} and Ga^{3+} into the zincian malachite precursor phase and a linear correlation of the (Zn,Ga) content in this phase with the catalytic activity of the final catalyst was observed.

Chapter 5: Promoting Methanol Synthesis Catalysts:
Correlations between Microstructure and Activity in Cu/ZnO/Ga$_2$O$_3$

5.1 Introduction

Cu/ZnO/(Al$_2$O$_3$) catalysts are of major industrial interest as they have been successfully applied in methanol synthesis for over 40 years. Furthermore, they are active in methanol steam reforming and water-gas-shift reaction. The synthesis route for preparing Cu/ZnO/(Al$_2$O$_3$) catalysts follows a multi-step procedure including temperature and pH-controlled co-precipitation of aqueous Cu,Zn,Al nitrate solution with sodium carbonate solution, aging, washing, drying, calcination and finally activation by reduction [1].

Metallic copper which is present as nanoparticles is regarded as the active species [2]. But there is still no general agreement about the nature of the active sites in Cu/ZnO catalysts [3]. However, it is known that highly productive catalysts exhibit a high copper surface area [4]. This requires an optimal dispersion of the active copper phase but additionally, the "quality" of the active Cu surface area plays a decisive role. ZnO is known to act as a geometrical spacer and to increase the Cu dispersion [5]. Thus, it leads to higher exposition of the active surface to the reaction gas. Beyond this geometrical influence, ZnO was reported to cause synergetic effects due to the interface contact with the copper phase [6]. Proposed active species are Cu-Zn alloy formed during reduction [7], dissolved Cu$^+$ in ZnO [8] or electron rich Cu at Schottky-junctions [9]. In our recently published model of the active site of industrial methanol synthesis over Cu/ZnO/Al$_2$O$_3$, the synergetic effect is accounted for by strong metal support interaction (SMSI) [10], which has been observed in high-performance catalysts by HRTEM and XPS. SMSI between Cu and ZnO has previously been reported in literature and studied by Cu surface area determination [11], EXAFS [12] and IR spectroscopy of CO adsorption [13].

We showed that the intrinsic activity of the exposed Cu surface area scales with the abundance of stacking faults in Cu nanoparticles [10]. This correlation was rationalized by the generation of high energy sites at the surface at the positions, where the planar defect terminates. Also residual oxygen in Cu as a result of incomplete reduction might play a role for the defect structure of active Cu. In this context, the role of promoters like alumina is of interest. Saito et al. [14] reported that addition of metal oxide promoters can have different effects, first the increase of the Cu dispersion in the case of alumina or zirconia, secondly the improvement of the specific activity in the case of gallia and chromia. The authors claim that the latter feature is due to the optimization of the Cu$^+$/Cu^0 ratio on the Cu surface under reaction conditions. Furthermore, Al$_2$O$_3$ is known to enhance the thermal and performance stability [15-16].

We have recently found that an Al content of around 3 mol% is sufficient to obtain a beneficial promoting effect [17]. Al was introduced into the ZnO phase at Zn^{2+} sites in tetrahedral coordination. Alike ZnO, the function of the Al_2O_3 promoter was divided into a geometrical and a synergetic contribution. The former affects the Cu dispersion and leads to an increase of the Cu surface area. The latter promotes the intrinsic activity of Cu and was related to the incorporation of Al into the ZnO lattice and an influence onto the Cu/ZnO synergy. In this work, we report on the effects of gallia on Cu/ZnO to study, whether the role of the structural promoter Al_2O_3 can be generalized. Part of the data on gallia has already been published previously [17].

In the literature studies dealing with promoters in the Cu,ZnO system are often carried out with high fractions of up to 25 mol% and more [14, 18-25]. Here, we concentrate on the region of smaller contents ($\leq$ 4 mol%) for which great promotion of catalytic performance has been observed for Al_2O_3 containing Cu/ZnO [17]. Additionally, some higher contents are applied for comparison. We report on introduction of gallia as promoter in the Cu,Zn system (Cu:Zn = 70:30). Aspects of precursor preparation and microstructural characterization as well as catalytic results in methanol synthesis are presented.

5.2   Experimental

5.2.1   Sample Preparation

Metal hydroxy carbonate precursors with fixed Cu:Zn ratio (70:30) and different Ga contents up to 13 mol% (metal base) were synthesized by co-precipitation from acidic Cu,Zn,Ga nitrate solutions and Na_2CO_3 solution as basic precipitating agent in an automated lab reactor (LabMax, Mettler Toledo) under controlled conditions like dosing, stirring, temperature (338 K) and pH value (6.5). After aging (60 min after pH drop) the slurry was filtered, the precipitate washed several times with water and spraydried (Niro minor mobile, T_{inlet} = 473 K, T_{outlet} = 373 K). Calcination was carried out in air at 603 K (2 K min^{-1}) for 3 h. The high Cu:Zn ratio of 70:30 is typically applied in industrial catalyst preparation and aims at a maximal incorporation of Zn into the zincian malachite precursor phase [26].

5.2.2 Sample Labeling

The designation of the samples was chosen with regard to the nominal Ga content, respectively. Ga4.0 means a nominal Ga content of 4.0 mol% of all metal species [Cu+Zn+Ga]. The designation Ga0.0 refers to the binary Cu,Zn reference sample without any trivalent promoter.

5.2.3 Elemental Analysis

All Cu,Zn,Ga calcined oxides were subjected to elemental analysis by XRF (Table 5-2: Results of calcined sample characterization and Cu surface areas). The determined Cu to Zn ratios were near the nominal ones of 70:30. The measured Ga contents (mol% based on metals) are slightly smaller than the nominal values. Probably, the deviations are due to the not exactly defined water content of the metal salts used for the synthesis. Most of all, Ga-nitrate is very hygroscopic.

5.2.4 Characterization

X-ray fluorescence spectroscopy (XRF) was performed after glassing the calcined samples with $Li_2B_4O_7$ in a Bruker S4 Pioneer X-ray spectrometer. X-ray diffraction (XRD) was applied to the catalyst precursors and calcined samples. The samples were measured on a STOE STADI P transmission diffractometer equipped with a primary focusing Ge monochromator (Cu $K_{\alpha 1}$ radiation) and a linear position sensitive detector (moving mode, step size 0.1 °, counting time 10 s/step, resolution 0.01 °, total accumulation time 634 s). The samples were mounted in the form of a clamped sandwich of small amounts of powder fixed with a small amount of grease between two layers of thin polyacetate film. The phase composition was determined by full pattern refinement in the 2θ range 4-80 ° according to the Rietveld method using the TOPAS software [27] and crystal structure data from the ICSD database. Specific surface areas were determined by N_2 physisorption in a Quantachrome Autosorb-6 machine after degassing the samples at 353 K for 2 h. Isotherms were recorded at liquid nitrogen temperature and evaluated according to the BET method. Thermogravimetric experiments (TGMS) were done on a NETZSCH Jupiter thermobalance in flowing air. The gas evolution was measured with a quadrupole mass spectrometer (Pfeiffer Vacuum, Omnistar).

Scanning electron microscopy (SEM) images were taken in a Hitachi S-4800 field emission gun (FEG) system. Transition electron microscopy (TEM) was performed with a Philips CM200FEG microscope operated at 200 kV and equipped with an EDX spectrometer. For TEM investigation, the samples were reduced up to a temperature of 523 K and transferred to the microscope in inert atmosphere. The coefficient of spherical aberration was Cs = 1.35 mm, and the information limit was better than 0.18 nm. High-resolution images with a pixel size of 0.016 nm were acquired at the magnification of 1083000x with a CCD camera, and selected areas were processed to obtain power spectra (square of the Fourier transform of the image), which were used for measuring interplanar distances and angles (accuracy ± 1% and ± 0.5 deg, correspondingly) for phase identification. Projected areas have been measured and equivalent diameters calculated for 1500-3000 Cu particles in each sample. In all cases the values of standard error of the mean diameter were ≤ 0.1 nm. Frequency distributions of the particle sizes fitted well to Lognormal functions. EDX analyses were performed for 5-15 larger aggregates containing at least several hundred particles in each sample.

Temperature programmed reduction (TPR) was performed with around 40 mg of each sample in a glass reactor, fixed by means of quartz wool plugs. The reduction was carried out in a CE instruments TPDRO 1100 machine with 80 mL min^{-1} 5% H_2 in Ar up to a temperature of 623 K (6 K min^{-1}). The K-values according to Monti and Baiker [28] were 100-120 s, the P-values according to Caballero [29] 10-12 K. The reduction progress was followed with an internal thermal conductivity detector. Analysis was performed with regard to the temperature with the highest H_2 consumption (T_{max}) and the total H_2 consumption with respect to the CuO content in the sample (compared with a pure CuO reference). In the following, the term "reducibility" is used for the latter feature. The CuO content of the samples was derived from XRF data with the assumption that only CuO, ZnO and Ga_2O_3 were present and under neglect of $HT-CO_3^{2-}$.

UV-Vis spectra were recorded in a Perkin-Elmer Lambda 650 High Performance Spectrometer equipped with a Harrick Praying Mantis diffuse reflectance attachment. The band gap energy (direct transition) was calculated by linear extrapolation of the function $[F(R_\infty)h\nu]^{1/2}$ versus $h\nu$ to 0, as suggested by Barton et al [30]. This procedure results from a linearization of the theory of direct and indirect band gap transitions in semiconductors.

X-ray absorption spectra were conducted of Ga K-edge (10.367 keV) at the X1 beamline at HASYLAB at DESY (Hamburg, Germany). Measurements were carried out in transmission mode. Intensities were detected in ion chambers before samples and behind samples and behind reference foils. With respect to maximal absorption of the samples at the Ga K-edge samples were diluted with wax and pressed with a force of 1 ton for 30 seconds to pellets with 13 mm in diameter. The spectra were calibrated to the position of K-edge of a reference foil (Zn foil for

Zn K-edge and Ga K-edge). Subsequently, background correction and normalization were performed. The software Athena 0.8.061 was used [31].

The copper surface area was determined by applying N_2O reactive frontal chromatography (N2O-RFC) based on the method proposed by Chinchen et al. [4]. Around 100 mg of a sieve fraction (100-200 μm) of each sample were placed in a stainless steel U-tube reactor and fixed by means of quartz wool plugs. The prior reduction was carried out in the same device and conditions as for TPR, but only up to a temperature of 523 K and with a holding time of 30 min. The reduction progress was additionally followed with a quadrupole mass spectrometer (Pfeiffer Vacuum, Omnistar). After cooling down to 303 K, the catalyst has been flushed for 30 min in pure Ar and 15 min in pure He in order to achieve an adsorbate-free reduced catalyst surface. N_2O-RFC was performed with 10 mL min^{-1} 1% N_2O in He, at which the N_2O reacts quantitatively with the Cu surface atoms forming gas-phase N_2. The specific Cu metal surface area has been calculated from the formed amount of N_2 using a value of $1.47*10^{19}$ atoms per m^2 for the mean Cu surface atom density. The error of the specific Cu surface area is about $\pm 1 \, m^2 g^{-1}$.

5.2.5 Catalytic testing

Catalytic testing in methanol synthesis was performed using an 8-channel parallel fixed bed reactor setup working at 60 bars of a synthesis gas mixture (59.5% H_2, 8.0% CO_2, 6.0% CO, rest inert). Gas analytics was done by gas chromatography. 200 mg of each catalyst sample (sieve fraction 100-200 μm) was loaded to the reactor and reduced prior to the measurement in diluted hydrogen at 523 K at ambient pressure. After a formation period of 48 hours on stream at 523 K, the catalytic activity was measured for 12 hours and stable conversions were detected. Afterwards, the reaction temperature was lowered to 483 K and the performance was measured again for 12 hours. The methanol weight time yields (WTY) were calculated using the methanol concentration in the outlet gas.

5.3 Results and Discussion

5.3.1 The influence of gallia on the precursor chemistry

Different mixed metal hydroxy carbonate precursor phases can emerge in the course of precursor preparation, notably $Cu_2(OH)_2CO_3$ (malachite) for pure Cu samples, $(Cu_{1-x}Zn_x)_2(OH)_2CO_3$ (zincian malachite) with $x < 0.3$, $(Cu_{1-y}Zn_y)_5(OH)_6(CO_3)_2$ (aurichalcite) with $y > 0.5$, and $(Cu,Zn)_6Al_2(OH)_{16}CO_3 \cdot 4H_2O$ (hydrotalcite-like phase), only when a significant amount of Al^{3+} is present. The last phase should also be formed with other trivalent ions instead of Al^{3+}, such as Ga^{3+}.

The critical role of the precursor chemistry has been emphasized in our earlier work [26]. All parameters applied in each single step of the catalyst preparation influence the bulk and surface structure and therewith the characteristics and activity of the resulting catalyst. This phenomenon is also called the "chemical memory" and means the influence of early stage parameters or rather the characteristics of the precursor phase on the microstructure and activity of the final catalyst [32-33].

For the industrially applied ternary system (Cu:Zn:Al = 60:30:10), a comprehensive study was accomplished by Baltes et al. wherein the influence of pH value and temperature during the precursor preparation was investigated [23]. For T = 343 K and pH 6-8, the best catalytic performance was achieved. A comprehensive understanding of the effect of addition of Al or Ga can only be achieved if all stages of catalyst preparation, precursor, calcined oxides and active catalyst, are carefully characterized.

5.3.1.1 XRD analysis

All precursor samples were subjected to XRD analysis for phase identification and determination of composition and crystal structures. Figure 5-1 shows the XRD patterns of selected precursors with 1.5, 2.5 and 3.5% nominal Ga content, respectively. Three different crystalline phases can easily be identified by selected not-overlapping reflections, namely (020) at 14.8 °2θ for zincian malachite, (400) at 13.0 °2θ for aurichalcite and (003) at 11.6 °2θ for a hydrotalcite-like phase.

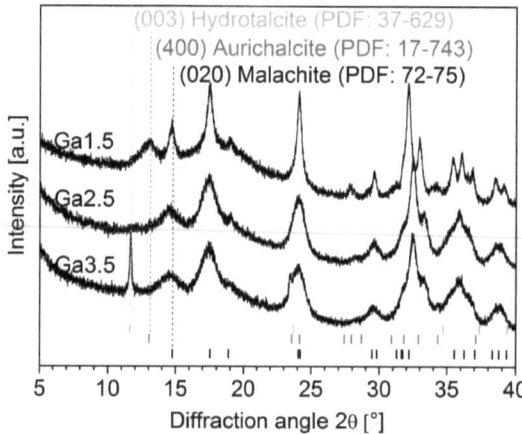

Figure 5-1: Selected XRD patterns of Cu,Zn,Ga hydroxy carbonate precursors with different Ga contents. Marks of the references malachite (black), aurichalcite (grey) and hydrotalcite-like phase (light grey) are included.

The phase compositions and domain sizes of the precursors were determined by Rietveld refinement. As representative example, the refinement result of Ga1.0 is depicted in Figure 5-2. It is noted that the accurate determination of the exact weight fractions in the phase mixture is difficult due to the low amounts of some components, the generally low crystallinity and the high noise of the XRD patterns. However, the general trends seen within the series of samples are regarded as reliable, while the absolute values of individual samples depend on the fitting constraints and have to be compared with care. Results of the full Ga series are shown in Figure 5-3. The binary Cu,Zn reference sample Ga0.0 contains about 88% zincian malachite and 12% aurichalcite as crystalline phases (Figure 5-3a). For increasing Ga content (Ga1.5), the fraction of aurichalcite is slightly increased up to 19% at the expense of zincian malachite. In Ga2.0, the zincian malachite fraction is heavily increased and reaches 100% in Ga2.5. But as mentioned, one has to be very careful with this statement, firstly because of the limits of this evaluation and secondly because it is only valid for crystalline phases which can be detected by XRD. Further increase of the Ga content leads to the formation of small but clearly detectable amounts (up to 2%) of a hydrotalcite-like phase. The phase fractions do not change significantly and stay nearly constant for Ga contents up to 13 mol%.

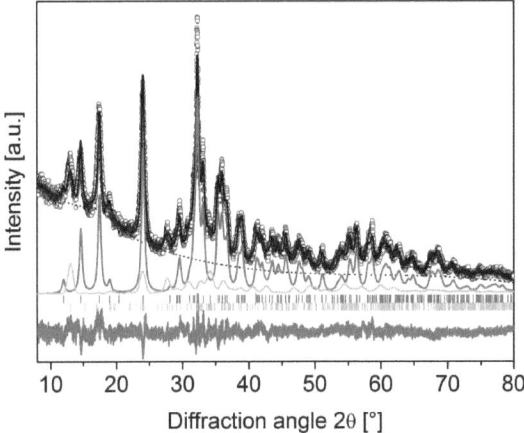

Figure 5-2: Rietveld refinement for the XRD pattern of a Cu,Zn,Ga hydroxy carbonate precursor sample containing 1.0% Ga, experimental data (open circles), total calculated curve (black), background (black, dotted), difference curve (grey), calculated pattern zincian malachite (grey), calculated pattern aurichalcite (light grey). The thick marks indicate the positions of the Bragg reflections. The fit quality is comparable for the fits of all other samples.

Also the domain size of zincian malachite can be extracted from the Rietveld refinement in form of the volume weighted mean column heights. As it can be assumed from the broadening of the XRD reflections (Figure 5-1), the crystallinity of this phase seems to decrease with increasing Ga content. Indeed, in the Ga series, the domain size (Figure 5-3b) steadily is decreasing from Ga0.0 (23.1 nm) to Ga4.0 (6.3 nm) and then remains constant. Only Ga1.5 does not follow this trend (22.4 nm). Analysis of a reproduced sample prepared at the same conditions confirms the result and indicates that this effect is real. The smallest domain size for zincian malachite is reached for a Ga content of 3 mol%. At higher Ga concentrations, only a low amount of hydrotalcite-like phase is formed and the zincian malachite stays nanocrystalline.

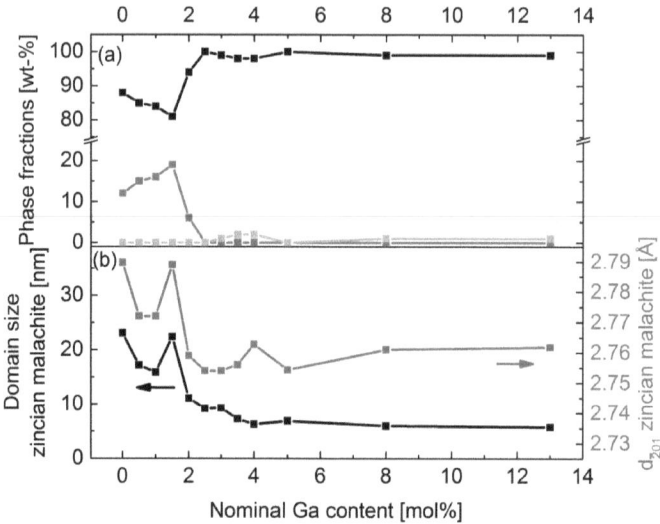

Figure 5-3: Results of XRD full pattern refinement of Cu,Zn,Ga hydroxy carbonate precursors. Top: phase fraction of zincian malachite (black), aurichalcite (grey) and hydrotalcite-like phase (light grey); Top: domain size of zincian malachite (black) and $d_{20\bar{1}}$ value (grey); errors of domain sizes are smaller than used symbols.

An important feature to estimate the potential of a catalyst is the position of the $20\bar{1}$ reflection of the precursor phase zincian malachite in the XRD patterns. It is correlated to the amount of Zn incorporation in this phase as was first reported by Porta et al. [34]. The $d_{20\bar{1}}$ value can be used as a quantitative measure of the incorporation of non Jahn-Teller-distorted ions like Zn^{2+}, Ga^{3+} (or Al^{3+}) into this phase. Low $d_{20\bar{1}}$ values correspond to high incorporation (the $d_{20\bar{1}}$ value of pure malachite, $Cu_2(OH)_2CO_3$, is about 2.863 Å). Compared to Ga0.0 (2.790 Å), the values become smaller for the Ga promoted samples (Figure 5-3b) and reach a minimum for Ga2.5 to Ga3.5 (2.754 Å). This value corresponds to a Zn incorporation of 31% assuming that the correlation of the binary system shown in [35-37] is also valid for the ternary precursors. Again, Ga1.5 delivers a runaway value. Ga4.0 and Ga5.0 show some scattering but in Ga8.0 and Ga13.0 the value seems to level off.

Remarkably, the domain (crystallite) sizes of zincian malachite and the Zn (Ga) content in zincian malachite (obtained from $d_{20\bar{1}}$ values) show a very similar qualitative trend within the Ga series (Figure 5-3b). Plotting these two features against each other delivers an approximately linear correlation (Figure 5-4). It can be concluded, that the degree of Zn (Ga) incorporation in the zincian malachite directly influence the crystallite size and can be controlled by the Ga content at a constant Cu:Zn ratio. An inverse correlation is found for the measured BET surface areas (Figure 5-4). For Ga0.0, the BET surface area is about 60 m² g⁻¹ and is heavily raised for increasing Ga contents up to 150 m² g⁻¹ for Ga3.0 whereupon a stable value of around 145 m² g⁻¹ is reached for higher Ga contents. Again Ga1.5 does not follow the trend.

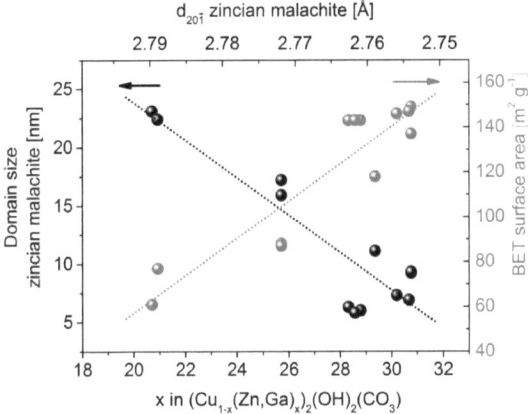

Figure 5-4: Zincian malachite domain sizes (black) and BET surface areas (grey) of Cu,Zn,Ga hydroxy carbonate precursors in dependence from Zn (Ga) incorporation into zincian malachite and $d_{20\bar{1}}$ value, respectively.

These results from XRD analysis of the precursors suggest that the limit for the Zn (Ga) incorporation into the zincian malachite structure is reached for a concentration of the promoter at around 2.5-4.0 mol% corresponding to a high phase fraction of zincian malachite and a low crystallite size. Beyond this concentration, the $d_{20\bar{1}}$ value cannot be further decreased. In contrast, it is slightly increased and settles down at a constant value probably due to competing incorporation of Zn and Ga in the hydrotalcite-like phase. Thus, low concentrations of Ga promote the incorporation of the Zn into zincian malachite. With increasing promoter content, the fraction of Zn-rich aurichalcite is reduced, leading to a decrease of the $d_{20\bar{1}}$ value indicating a higher Zn (Ga) substitution of zincian malachite. For promoter contents below 2.5 mol%, no crystalline phases originally containing M^{3+} ions were found, suggesting that either Ga^{3+} is present as an amorphous phase which cannot be detected by XRD, or introduced into the aurichalcite or the zincian malachite structure. Insertion into zincian malachite is known for Al^{3+} [17] and should be transferable to Ga^{3+} as well. In this case, the excess of positive charge by Ga^{3+} ions on Cu^{2+} or Zn^{2+} sites has to be compensated. Possible mechanisms are the formation of cation or proton vacancies. For Ga contents higher than 2.5 mol%, the M^{3+} ions probably cannot be further inserted into the zincian malachite phase due to the charge mismatch.

Unlike the results for Al as promoter [17], the fraction of the hydrotalcite-like phase stays surprisingly stable at only 2 wt%. No GaOOH was detected. This indicated, that at high Ga contents, a Ga containing phase is formed which is X-ray amorphous. Using Al as a promoter, large fractions of a hydrotalcite-like phase also have been observed when exceeding a critical promoter content [17]. The critical promoter contents for high incorporation of Zn and Ga in zincian malachite and a large fraction of this phase are in the range of 2.5-4.0 mol% for both, Ga and Al.

5.3.1.2 Scanning electron microscopy

The precursor sample Ga13.0 was investigated with SEM to study the Ga distribution and find evidence for a segregated Ga-rich phase. As expected, the sample is inhomogeneous. Figure 5-5a shows an overview about a typical region of the sample. The backscattered electron microscopy image of the same area delivers a mean atomic number contrast (Figure 5-5b) and shows a brighter area which should correspond to a higher mean atomic number.

Altogether, the predominantly observed morphology of the Cu,Zn,Ga precursor in spot 1 (Figure 5-5c) is consistent with an earlier study on the Cu,Zn,Al system, where zincian malachite was described as needles of 20 nm × 200 nm [26]. Energy-dispersive X-ray spectroscopy (Figure 5-5e) of six different regions with the mentioned morphology showed an average local elemental composition of 64 ± 2% (Cu), 25 ± 1% (Zn) and 11 ± 2% (Ga) which agrees to the results from XRF in Table 5-1 (63:25:12).

But aside this finding, some smaller regions with different morphology, e.g. platelets or particles, were detected. In spot 2 (Figure 5-5d), no needles were observed but the particle looks sponge-like. The local composition showed high Ga contents at the expense of Cu which explains the brighter area of spot 2 in Figure 5-5b. The Ga-richest composition was 33:26:41 where the M^{2+} to M^{3+} ratio ($\approx$ 3:2) does not fit to that of a hydrotalcite-like phase (3:1). Since EDX cannot be applied to infinitely small regions, it cannot be finally clarified whether this is a matter of a single phase or a superposition of a Cu rich phase (zincian malachite) and an unidentified Ga rich phase. Interestingly, the Zn content in all measured regions was constant within 24-28 mol% and thus independent of the Cu and Ga fraction.

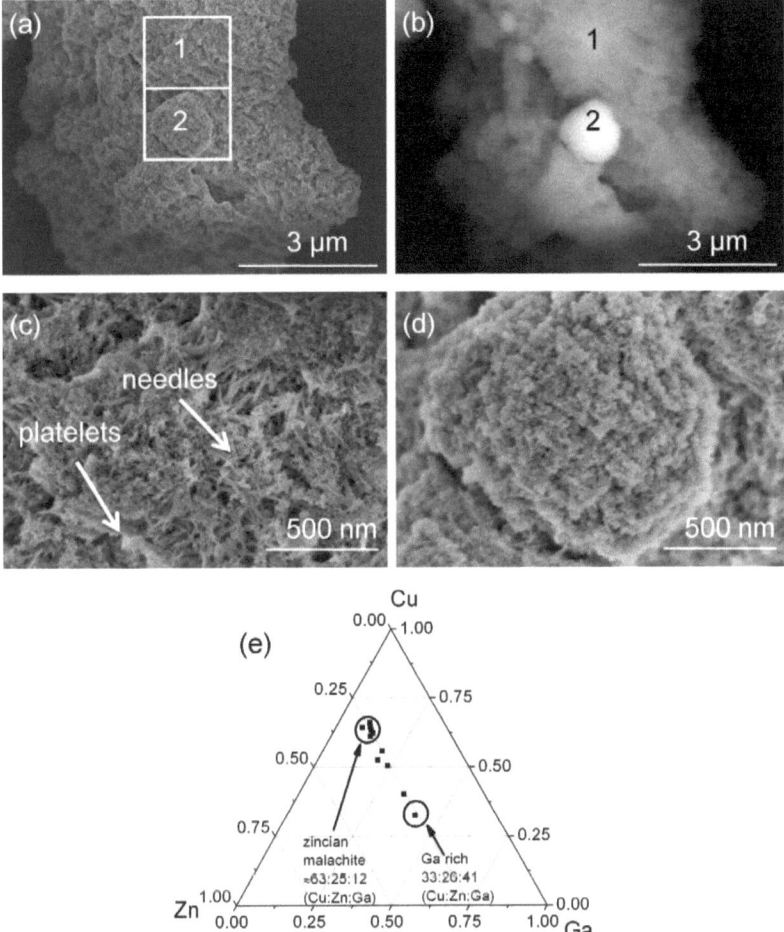

Figure 5-5: Electron microscopy images (2.5 keV) and EDX-results of the Ga13.0 hydroxy carbonate precursor showing an overview (a), the same overview in backscattered electron mode (b), zincian malachite needles and some platelets (c), Ga-rich area (d) and local elemental distribution from SEM-EDX (e).

Chapter 5: Promoting Methanol Synthesis Catalysts:
Correlations between Microstructure and Activity in Cu/ZnO/Ga$_2$O$_3$

Table 5-1: Results of precursor sample characterization

Label	Sample number (internal)	BET surface area [m^{2}g^{-1}]	Phase composition [wt%]			Domain size zincian malachite [nm]	$d_{20\bar{1}}$ [Å]	Thermal Analysis		
			Zincian malachite	Aurichalcite	Hydrotalcite-like phase			Mass loss [wt%]	T_{max}[a] [K]	HT-CO$_2$[b] [%]
Ga0.0	7399	61	88	12	0	23.1	2.790	27.7	723	52
Ga0.5	10220	88	85	15	0	17.2	2.772	28.4	747	50
Ga1.0	10210	87	84	16	0	15.9	2.772	28.3	760	50
Ga1.5	10222	77	81	19	0	22.4	2.789	28.1	756	52
Ga2.0	10216	118	94	6	0	11.1	2.759	28.3	775	49
Ga2.5	10224	137	100	0	0	9.2	2.754	28.4	789	45
Ga3.0	10212	149	99	0	1	9.3	2.754	28.8	776	45
Ga3.5	10482	146	98	0	2	7.3	2.756	28.3	788	46
Ga4.0	12045	143	98	0	2	6.3	2.763	28.1	785	37
Ga5.0	12063	147	100	0	0	6.9	2.754	27.4	792	39
Ga8.0	12079	143	99	0	1	6	2.761	26.8	783	38
Ga13.0	12158	143	99	0	1	5.8	2.762	26.5	781	31

[a] Temperature of maximum CO$_2$ emission
[b] CO$_2$ emission above 673 K relative to overall CO$_2$ emission according to MS signal

5.3.1.3 Thermal analysis

Thermal decomposition of the precursor is necessary to obtain nano-sized CuO and ZnO particles. This calcination step can be followed by thermogravimetric measurement (TG) combined with evolved gas analysis (EGA). In contrast to the calcination, which is performed up to a temperature of 603 K, the TG-EGA experiments were executed from 303-973 K. The anions in the Cu,Zn,X hydroxy carbonate are decomposed under emission of water and carbon dioxide. Exemplarily, the results for Ga2.5 are depicted in Figure 5-6, showing the development of the mass loss and the normalized H$_2$O and CO$_2$ traces. XRD patterns of the samples thermally treated at 973 K revealed crystalline CuO and ZnO (not shown). The domain sizes of both phases decreased by a factor of around three within the Ga series indicating the geometrical effect of the promoter. The XRD patterns of Ga8.0 and Ga13.0 showed weak and broad reflections of ZnGa$_2$O$_4$.

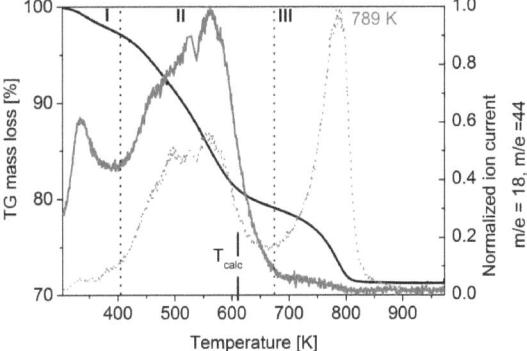

Figure 5-6: TG-MS results of Ga2.5 hydroxy carbonate precursor: mass loss (black), MS traces of H_2O (grey) and CO_2 (grey, dotted).

The theoretical mass losses for pure zincian malachite $(Cu_{1-x}Zn_x)_2(OH)_2CO_3$ and aurichalcite $(Cu_{1-y}Zn_y)_5(OH)_6(CO_3)_2$ account for 28% and 26%, respectively, and show only a slight dependence on the Cu:Zn ratio because of the similar molar masses of Cu and Zn. However, the Ga incorporation and charge compensation effects in these phases as well as the varying amount of incorporated or physisorbed water were neglected during this calculation. The theoretical mass loss for the hydrotalcite-like phase $(Cu_{1-z}Zn_z)_6(Ga)_2(OH)_{16}CO_3 \cdot 4H_2O$ is 20%. Thus, the highest mass loss is expected for pure zincian malachite. Indeed, Figure 5-7a shows the highest mass loss of 28.8% for the sample Ga3.0, which is in the regime of phase-pure zincian malachite according to XRD. For Ga contents higher than 3 mol%, the mass loss is monotonically decreasing down to 26.5% (Ga13.0). EGA shows that the decomposition of Cu,Zn,Ga hydroxy carbonates mainly proceeds in three steps (Figure 5-6). After the release of physisorbed, incorporated or interlayer water (range I, up to 403 K), the second step is characterized by simultaneous emission of H_2O and CO_2 (range II, ca. 403-673 K). In the third step, only CO_2 is emitted at high temperatures (range III, ca. 673-873 K). The origin of this last decomposition step is the presence of temperature stable carbonate species ($HT-CO_3^{2-}$) which are probably located at the interface between the formed CuO and ZnO [33, 38]. The role of this residual carbonate, which is still present after calcination at not too high temperatures, is debated and it has been proposed that it can stabilize oxidized copper species in the reduced catalyst and increase the activity [39]. The abundance and the stability of these species are suggested as measures for the amount and the quality of the interfaces and grain boundaries of CuO/ZnO aggregates, respectively. Accordingly, pure malachite $Cu_2(OH)_2CO_3$ and hydrozincite $Zn_5(OH)_6(CO_3)_2$ do not show the emission of these species [33, 36, 38].

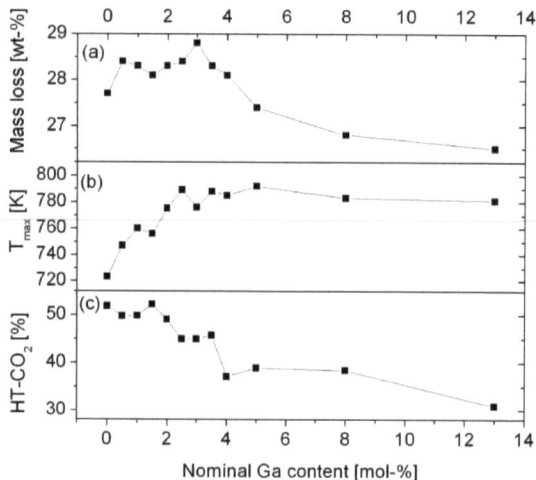

Figure 5-7: TG-MS results of Cu,Zn,Ga hydroxy carbonate precursors: mass loss after heating to 973 K (a), temperature of highest CO_2 emission rate (b) and CO_2 emission above 673 K relative to overall CO_2 emission according to MS trace (c).

The HT-CO_3^{2-} decomposition temperature first increases with Ga content from 723 (Ga0.0) up to a temperature of 792 K (Ga2.5) and then stays nearly constant indicating very stable HT-CO_2 (Figure 5-7b). This indicates stronger interaction of CuO and ZnO with higher Zn (Ga) incorporation into zincian malachite.

The HT-CO_3^{2-} amount can be calculated from the fraction of HT-CO_2 related to the overall CO_2 emission in a semi-quantitative manner. This fraction is relatively constant around 50% for Ga0.0 to Ga2.0 and then starts to diminish down to 31% in the Ga series (Figure 5-7c). Thus, the HT-CO_3^{2-} cannot be intrinsic to the unidentified Ga-phase but originates from the synthetic Cu,Zn hydroxy carbonate. The decrease rather suggests that the Ga-rich by-phase contains loosely bound carbonate itself and contributes to the CO_2 emission in range II (Figure 5-6).
These results show that the effect of the Ga promoter on precursor chemistry is also reflected in the thermal properties of the carbonate, which is still present in the catalyst after calcination.

5.3.2 Calcined samples

5.3.2.1 XRD analysis

Exemplarily, the XRD pattern of Ga2.5 after calcination is shown in Figure 5-8. All patterns were analyzed by Rietveld refinement (not shown), revealing CuO as the main phase and only small amounts (up to 1 wt%) of crystalline ZnO. Crystalline $ZnGa_2O_4$ spinel, which is commonly formed during calcination of hydrotalcite phase, was not found. Due to the homogeneous metal distribution in the precursor, a good dispersion of CuO and ZnO should be achieved. The CuO domain sizes are listed in Table 5-2. For Ga0.0 it is 4.7 nm and decreases down to 2.9 nm in the region between Ga2.0 and Ga5.0 where it reaches a minimum of 2.4 nm for Ga3.0. Higher Ga contents do not much affect the CuO domain size. A roughly inverse trend is observed for the BET surface areas (Table 5-2). This is expected because small CuO domains are a result of better nanostructuring which leads to higher BET surface areas.

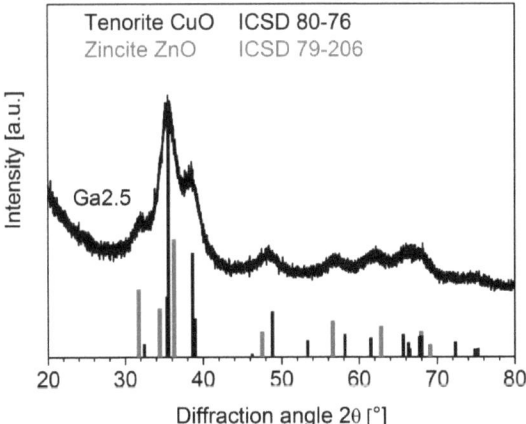

Figure 5-8: XRD pattern of calcined Ga2.5 sample.

Table 5-2: Results of calcined sample characterization and Cu surface areas

Label	Sample number (internal)	Ga-content [mol%] (XRF)	Cu:Zn ratio (XRF)	BET surface area [m² g⁻¹]	Domain size CuO [nm]	TPR results T_{max}^a [K]	TPR results H_2 cons.^{b} [%]	Cu surface area [m² g$_{CuO}^{-1}$]
Ga0.0	7400	0	71:29	110	4.7	481	101	25.3
Ga0.5	10221	0.4	71:29	96	5.3	474	104	38.4
Ga1.0	10211	0.8	71:29	97	3.9	476	95	43.4
Ga1.5	10223	1.2	71:29	91	5.3	475	101	38.2
Ga2.0	10217	1.5	70:30	117	2.9	479	96	47.9
Ga2.5	10225	2	71:29	117	3.3	480	101	48.1
Ga3.0	10213	2.5	71:29	108	2.4	475	96	51.5
Ga3.5	10483	2.9	71:29	115	3.7	474	101	45.7
Ga4.0	12046	3.5	70:30	117	3.3	474	103	51.1
Ga5.0	12065	4.4	71:29	131	2.7	483	100	45.9
Ga8.0	12081	7.2	71:29	118	3.4	480	100	50.6
Ga13.0	12159	12	71:29	126	3.2	476	97	47.8

^{a} Temperature of the highest H_2 consumption rate according to TCD signal
^{b} H_2 consumption relative to the amount of CuO contained in the sample

5.3.2.2 The influence of gallia on ZnO

To gain more insight into the state of the Ga promoter in the samples, K-edge X-ray absorption spectroscopy was applied for selected samples and five oxidic Ga references (α-gallia, β-gallia, γ-gallia, $ZnGa_2O_4$ and ZnO doped with 3 mol% of Ga). To confirm that Ga is incorporated in ZnO in the ZnO/3%Ga reference sample, UV-Vis measurements were performed with pure ZnO and ZnO/3%Ga in order to confirm the change of the optic properties due to doping. The band gap energy (direct transition) of pure ZnO (3.28 eV) was decreased to 3.13 eV for ZnO/3%Ga (Figure 5-9). This band gap energy reduction was already seen by eye, because ZnO/3%Ga has a yellow color whereas gallia is white and ZnO only has a slightly yellow touch. Change of optic properties by doping ZnO with small amounts of Ga or Al has been reported in literature [40-43] and naturally is accompanied also by the change of electrical properties. The calcined sample Ga2.5 was chosen because it was derived from a phase pure precursor material and should not contain large amounts of segregated Ga-phases. However, the results were found to be similar to those of Ga3.5 reported in ref. [17], whose precursor was not phase-pure.

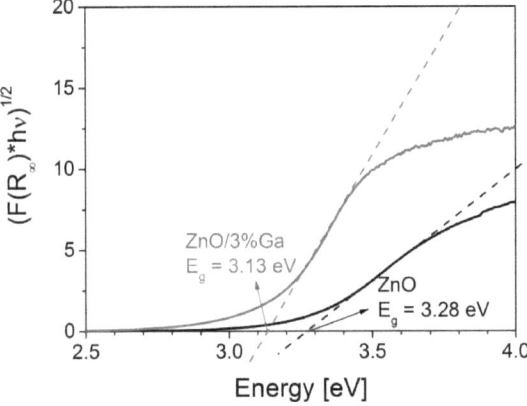

Figure 5-9: Determination of absorption edge energies for pure ZnO and Ga-doped ZnO from UV-Vis measurements by the intercept of a linear fit.

The experimental Ga-K-XANES spectrum of Ga2.5 was simulated as a linear combination of oxidic Ga reference spectra using the software Athena 0.8.061 [31] according to the method of least squares fit. The fitting procedure was applied in the fitting region of -20 to 50 eV (related to the Ga-K-edge of 10367 eV) for all possible 21 combinations of the five references. γ-gallia seemed not to contribute and was excluded. The results of the remaining 11 fits (Table 5-3) were sorted by fit quality, represented as R-values, whereas the lowest R-value refers to the best fit. The E_0-shifts were smaller than 0.3 eV in all cases. The fit results show that in principle each of the remaining Ga oxide references (α-gallia, β-gallia, $ZnGa_2O_4$, ZnO doped with 3 mol% Ga) can be present in the sample Ga2.5 because the R-values of fits 1-4 are similar. But it is apparent that no satisfying fit is possible without using the Ga-doped ZnO and $ZnGa_2O_4$ spinel reference. Especially simulating the region around 10388 eV needed the contribution of this reference which is visible in Figure 5-10 showing the best linear combination fit.

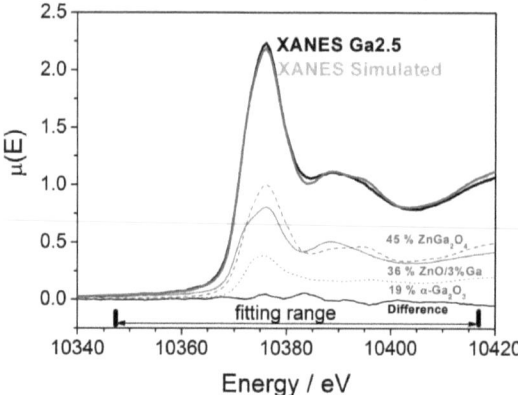

Figure 5-10: Ga K-edge XANES of the calcined Ga2.5 sample (black) and results of the linear combination fit in the range of -20 to 50 eV (related to the Ga-K-edge) using experimental spectra of Ga oxide reference materials.

Table 5-3: Results of linear combination fit of the Ga-K-edge XANES of the calcined Ga2.5 sample by Ga-oxide-reference spectra (the lowest R-value corresponds to the best fit).

Fit-Nr.	R-value (*10^{-3})	α-Ga$_2$O$_3$	β-Ga$_2$O$_3$	ZnGa$_2$O$_4$	ZnO/3%Ga
1	0.50	19%	0%	45%	36%
2	0.50	19%	-	45%	36%
3	0.55	-	13%	59%	29%
4	0.64	-	-	64%	36%
5	1.11	52%	-	-	48%
6	1.11	52%	0%	-	48%
7	1.20	-	35%	65%	-
8	1.20	0%	35%	65%	-
9	2.49	14%	-	86%	-
10	3.76	-	47%	-	53%
11	4.08	51%	49%	-	-

In agreement with the results reported in ref.[17], the Ga promoter in the calcined samples seems to be present in different oxidic species, one of them Ga incorporated into ZnO where Ga^{3+} is tetrahedrally coordinated by oxygen. Hence, Ga is not only increasing the Zn incorporation into the zincian malachite precursor but also modifies the ZnO component in the resulting catalyst by partial substitution of Zn^{2+} with Ga^{3+}. This leads to a change of different properties, like redox behavior, electronic structure and defect chemistry. The redox properties are changed as it can be seen from a comparison of TPR profiles whereas pure ZnO shows no signal but in the case of ZnO doped with 3 mol% Ga probably a small fraction of ZnO is reduced (Figure 5-11). The change of the electronic structure is obvious from the different band gaps (Figure 5-9). A modified defect chemistry of doped ZnO has been reported in literature [44-45] and is expected to be present in our Ga containing samples as well. Altogether, the changed properties will affect the intrinsic activity of the Cu/Zn(Ga)O probably via modified Cu-ZnO synergy.

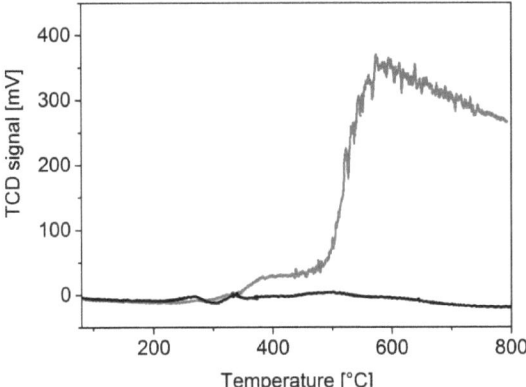

Figure 5-11: TPR profiles of ZnO (black) and ZnO/3%Ga (grey).

5.3.2.3 Temperature programmed reduction

The reduction behavior of the calcined samples was investigated by temperature programmed reduction (TPR) in hydrogen. It was not possible to describe the reduction profiles with a single peak because at least two shoulders were observed (Figure 5-12). Reasons might be reduction of CuO in multiple steps [46], reduction of multiple CuO species, e.g. from different precursor phases [47], or reduction of other components than CuO. The last two possibilities seem unlikely at least for the catalyst derived for phase pure zincian malachite precursors.

Reduction profiles were analyzed and the results are summarized in Table 5-2. No big change or clear trend were observed neither for the temperature of the highest H_2 consumption (474-483 K) nor for the reducibility of CuO (95-104% ± 5%). This is a striking example that the careful characterization of the early preparation stages may reveal a much greater wealth of information compared to the later steps of Cu/ZnO catalyst synthesis.

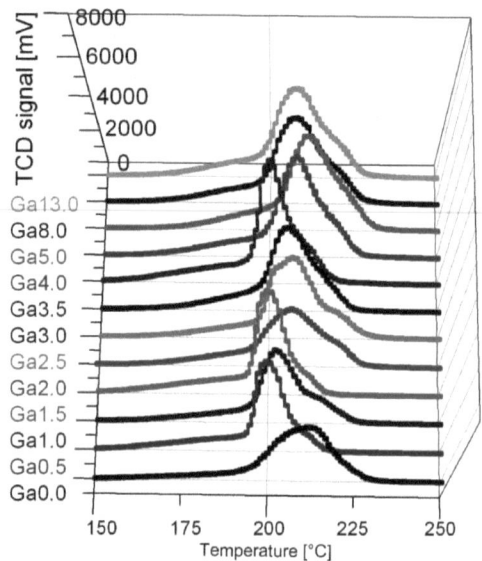

Figure 5-12: TPR profiles of Cu,Zn,Ga calcined samples.

5.3.3 Activated samples

5.3.3.1 Transmission electron microscopy

Selected reduced samples were subjected to TEM analysis: Ga0.0 as reference, Ga2.5, the sample with the lowest $d_{20\bar{1}}$ value and Ga13.0. In all cases a porous arrangement of roundish copper particles separated by zinc oxide particles was found indicating that the nanoparticulate microstructure of the calcined samples was conserved after reduction (Figure 5-13). The Cu particles sizes were around 11.1 nm (Table 5-4). The lowest value was found for Ga2.5. The local elemental composition was determined with TEM-EDX at different locations of the samples. The average results are shown in Table 5-4 and show a good agreement with the composition obtained by XRF (Table 5-2). Nevertheless, Ga13.0 shows a conspicuously large standard deviation for elemental distribution of Cu and Ga, which is in agreement with the findings from SEM of the precursor. From a triangular TEM-EDX composition diagram it becomes apparent that there is a positive correlation of the Zn and the Ga contents (Figure 5-14a; dotted line). For an independently varying fraction of a pure Ga and a "binary" Cu/Zn (70:30) component, a negative correlation would be expected (full line). The data points can be extrapolated to a Zn content of 33% on the binary Zn-Ga line, which equals the spinel composition. It can be concluded that the elemental composition of the individual single spots is a superposition of two different phases, the first originating from zincian malachite (Cu:Zn = 70:30) and the second is $ZnGa_2O_4$ spinel (Zn:Ga = 67:33). The latter was observed as small crystallites on the Cu particles (Figure 5-14b). The formation of $ZnGa_2O_4$ by solid state reaction from separated ZnO and gallia is unlikely at the low temperatures applied. Although the trend of the local elemental distribution in the precursor sample (Figure 5-5e) is not that pronounced with respect to the tendency of spinel composition, it is quite probable that the precursor of the spinel was an amorphous Zn,Ga hydrotalcite-like phase what would answer the question of the unidentified precursor phase.

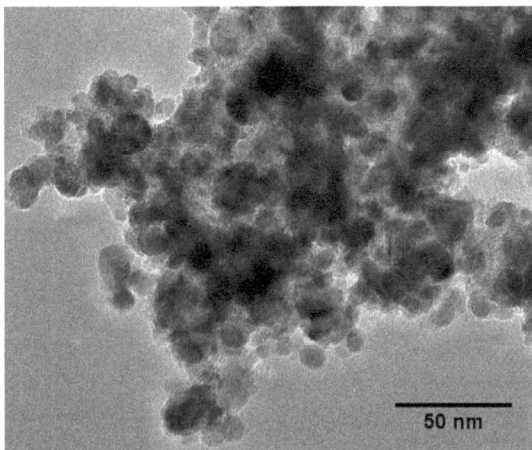

Figure 5-13: TEM image of the reduced Ga2.5 sample showing the typical arrangement.

Table 5-4: Results of reduced sample characterization and activity measurements

Label	TEM results				Interface ratio of Cu particles [%]	MeOH productivity (relative) [%]	Intrinsic activity (relative) [%]
	TEM-EDX [mol%]			Cu particle size [nm]			
	Cu	Zn	Ga				
Ga0.0	69.8 (2.0)	30.2 (2.0)	0	11.1 (4.5)	41	100	100
Ga1.0	-	-	-	-	-	134	80
Ga2.0	-	-	-	-	-	151	82
Ga2.5	67.6 (1.6)	30.0 (1.7)	2.4 (0.1)	10.5 (4.3)	19	165	90
Ga3.0	-	-	-	-	-	162	83
Ga3.5	-	-	-	-	-	163	95
Ga13.0	59.7 (9.2)	25.1 (1.7)	15.2 (4.4)	11.5 (4.4)	11	-	-

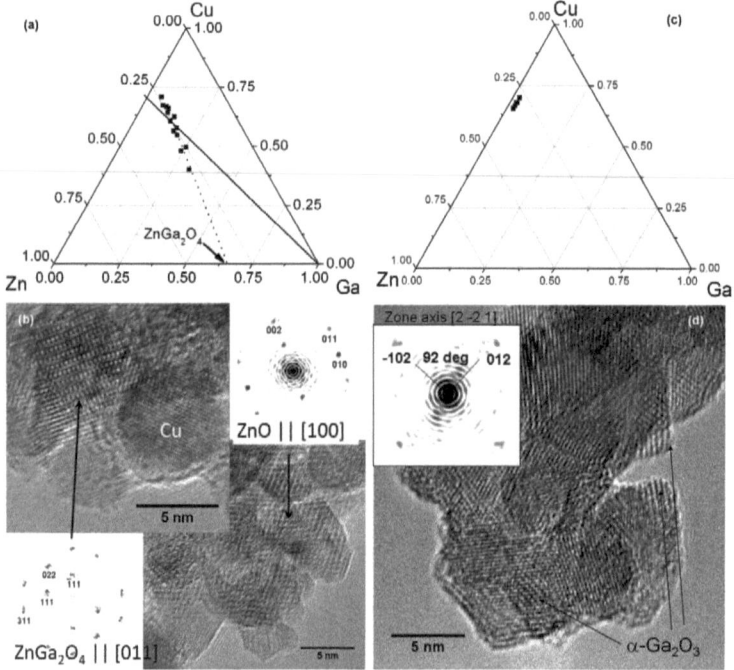

Figure 5-14: Results from transmission electron microscopy of reduced Ga,Zn,Ga samples. Ga13.0: elemental composition from TEM-EDX showing a strong tendency of $ZnGa_2O_4$ spinel formation (a) and TEM image of Cu particles covered with $ZnGa_2O_4$ spinel and ZnO (b). Ga2.5: elemental composition from TEM-EDX showing homogeneous distribution (c) and TEM image of crystalline α-gallia (d).

In contrast, TEM investigation of Ga2.5 showed homogeneous local elemental distribution (Figure 5-14c) and images with some small crystallites of α-gallia (Figure 5-14d) in agreement with the XANES results where α-gallia was found to be present according to the two best fits. It is assumed that this phase is formed during precursor calcination by segregation. Another possibility is the formation of α-gallia during reduction which can lead to a separation of Cu and Ga. No crystalline $ZnGa_2O_4$ spinel was found in HRTEM images.

5.3.3.2 Cu surface areas

Because the overall Cu content in the reduced samples slightly decreases with increasing Ga content, Cu surface areas from N_2O-RFC were calculated for a better comparison in relation to the contained CuO before reduction based on XRF under assumption that only CuO, ZnO and Gallia were present. A pronounced difference of the Cu surface areas of Ga0.0 (25.3 $m^2\,g_{CuO}^{-1}$) and Ga0.5 (38.4 $m^2\,g_{CuO}^{-1}$) can be observed in Figure 5-15a. Hence, already small amounts of Ga improve the Cu dispersion or the gas accessibility. With increasing Ga contents up to 3 mol%, the Cu surface area is increased up to 52 $m^2\,g_{CuO}^{-1}$. For Ga contents higher than 3 mol%, the values scatter in the region between 46 and 51 $m^2\,g_{CuO}^{-1}$. With regard to the increased Cu surface areas of Ga2.5 and Ga13.0 compared to Ga0.0 at a very similar particle size (from TEM), the effect of increased Cu surface area can be explained by less embedment of the Cu particles in the oxidic phase which results in better gas accessibility due to increased porosity. A correlation between Cu surface areas and the CuO domain sizes was found for the Ga containing samples (Figure 5-15b).

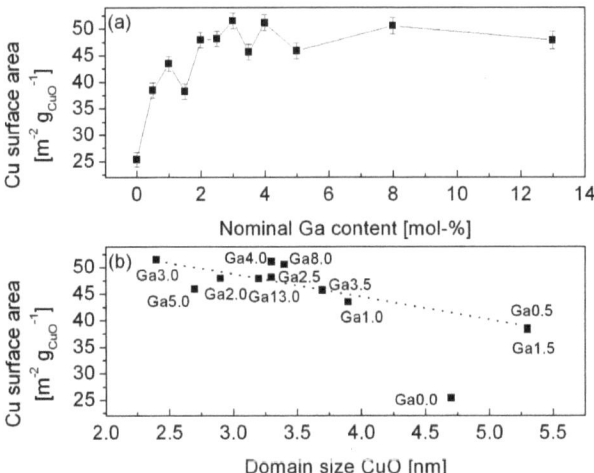

Figure 5-15: Cu surface areas of reduced Cu,Zn,Ga samples with respect to the calcined sample mass (a) and contained CuO mass (b). The error was estimated to be ± 1 $m^2\,g^{-1}$ in the top graph. CuO domain sizes of the calcined samples are given for comparison (c). Errors for domain sizes were smaller than the used symbols.

From the TEM Cu particle size and known Cu content, a theoretical Cu surface area (assuming isolated roundish particles) was calculated. Comparison with experimentally obtained Cu surface areas from N_2O-RFC delivered the interface ratio of Cu particles and indicated their degree of embedment. Therein, contact with other Cu particles or the oxidic matrix is possible. Although the absolute values are very inaccurate, the trend of decreasing interface ratio (equals decreased embedment) with increasing Ga content is discernable (Table 5-4).

5.3.4 Methanol synthesis activity

The activity in methanol synthesis was measured for selected Ga samples. The promoting effect of Ga on the catalytic performance is clearly visible: Starting from Ga0.0 (Figure 5-16), the activity increased and reached a maximum at Ga2.5 which is 60% higher than the value for the unpromoted sample Ga0.0. After this, the activity did not change significantly for Ga contents up to 3.5 mol%.

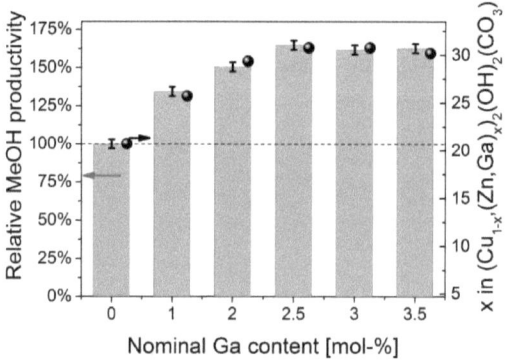

Figure 5-16: Methanol productivity of selected Cu/ZnO/Ga$_2$O$_3$ catalysts in methanol synthesis relative to the unpromoted sample Ga0.0 (grey bars) and content of non-Jahn-Teller ions (Zn^{2+} and Ga^{3+}) in the zincian malachite structure (black spheres) calculated from d$_{20\bar{1}}$ values of the binary Cu,Zn system according to [35].

The corresponding d$_{20\bar{1}}$ values of the precursors are converted into contents of non-Jahn-Teller-ions (Zn^{2+} and Ga^{3+}) in the zincian malachite according to the correlation found for the binary Cu,Zn system [35]. These values are added in Figure 5-16 and show the same trend like the activities so that a linear correlation can be assumed.

The correlation between activity and Cu surface area (not shown), however, shows little but significant scattering. This can be explained by different intrinsic activities of the exposed Cu surface areas, calculated by dividing the activity by the Cu surface area (Figure 5-17). The highest intrinsic activity is found for Ga0.0. In principle, the density of catalytically active sites on the surface of the Cu particles is described and different intrinsic activities have often been reported for different Cu based catalytic systems [11, 48].

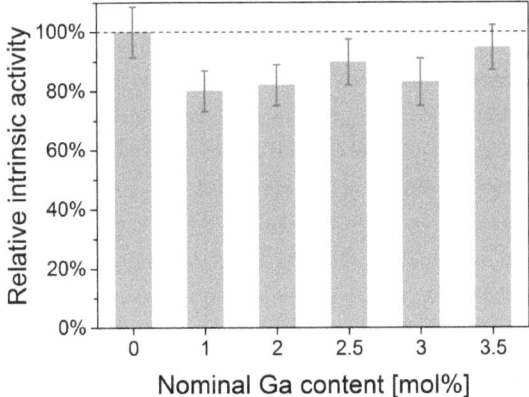

Figure 5-17: Intrinsic activities (related to the Cu surface area) of Cu/ZnO/Ga$_2$O$_3$ catalysts in methanol synthesis relative to the unpromoted sample Ga0.0.

In the case of Ga, the promoting effect, which leads to increased activity, can only be explained by the strong increase of the accessible Cu surface area up to 52 m^2 g$_{CuO}^{-1}$ (Ga3.0) compared to 25 m^2 g$_{CuO}^{-1}$ (Ga0.0) which means a doubling. This geometrical effect is promoted by Cu^{2+} dilution in the zincian malachite precursor with non-reducible cations. On the other hand, the overall activity is only improved by 60% (Ga3.0) which implies a lower intrinsic activity compared to Ga0.0. This effect is probably related to the modification of the ZnO phase by incorporated Ga which might have a negative effect on Cu-ZnO synergy. Another explanation would be that the intrinsic activity is related to the degree of embedment. Less embedment means also less contact to the ZnO phase. The generation of active centers which require the presence of ZnO is probably reduced.

The measured activity has to be regarded as a convolution of Cu surface area and intrinsic activity. In this sample series, the resulting improved activity due to Ga promotion is achieved due to a beneficial geometrical dispersion effect although the intrinsic activity is lowered. Both, geometrical and synergetic function of Ga promotion can be traced back to incorporation of Ga^{3+} in the zincian malachite precursor, explaining the correlation of $d_{20\bar{1}}$ and activity data.

5.4 Conclusions

In summary, it was shown that Ga^{3+}, similar to Al^{3+}, has a promoting effect on Cu/ZnO methanol synthesis catalysts. The promoting effect is most effective if only low amounts of Ga

– around 3 mol% – are added to the preparation. Higher Ga loadings lead to a Ga segregation and formation of amorphous $ZnGa_2O_4$ spinel in the final catalyst. Thus, the microstructure of the catalyst becomes inhomogeneous with Ga-rich domains of presumably lower catalytic activity. Low amounts of Ga^{3+} can be incorporated into the zincian malachite precursor phase and lead to an increase in methanol synthesis activity of 60% compared to the binary Ga free system. The promoting effect is mainly traced back to a geometric contribution: The Cu dispersion is increased by Ga due to better dilution of Cu^{2+} in the precursor by Zn^{2+} and Ga^{3+}, leading to a better interdispersion of metallic and oxidic components and a higher porosity of the resulting Cu/ZnO aggregates. This effect leads to an increase of the accessible Cu surface area by 100%. Apparently, the presence of Ga slightly diminishes the intrinsic activity of the exposed Cu surface area . This effect is probably related to the fraction of Ga species that are found to be incorporated into the ZnO phase after calcination of the precursor, leading to a modification of the properties of ZnO associated with a negative effect on the well-known Cu-ZnO synergy. The lack of synergy might also be enhanced by less embedment of the Cu particles leading to less interface contact with ZnO. The functionality of Ga promotion depends critically on the homogeneous distribution of Ga. The best distribution is achieved by incorporation into the zincian malachite precursor phase and a linear correlation of the (Zn,Ga) content in this phase with the catalytic activity of the final catalyst was observed.

Acknowledgement

Frank Girgsdies (help with XRD pattern analysis), Edith Kitzelmann (XRD measurements), Achim Klein-Hoffmann and Olaf Timpe (XRF), Gisela Lorenz (BET measurements), Genka Tzolova-Müller (UV-Vis), Antje Ota and Liandi Li for providing the Ga oxide reference samples, Juliane Scholz, Alexander Müller and Anke Walter for help with XAS measurement are acknowledged. Stefanie Kühl is acknowledged for valuable discussions, HASYLAB/DESY (Hamburg) for allocation of beamtime. Financial support was given by the German Federal Ministry of Education and Research (BMBF, FKZ 01RI0529, 2005-2008) and the STMWFK (NW-0810-0002, since 2010). Robert Schlögl is greatly acknowledged for valuable discussions and his continuous support.

5.5 References

[1] D. Waller, D. Stirling, F. S. Stone, M. S. Spencer, *Faraday Discuss.* **1989**, *87*, 107-120.
[2] K. C. Waugh, *Catal. Today* **1992**, *15*, 51-75.
[3] J. B. Hansen, P. E. H. Nielsen, in *Handbook of Heterogeneous Catalysis, Vol. 6* (Eds.: G. Ertl, G. Knözinger, F. Schüth, J. Weitkamp), Wiley-VCH, Weinheim 2nd ed., **2008**, pp. 2920-2949.
[4] G. C. Chinchen, K. C. Waugh, D. A. Whan, *Appl. Catal.* **1986**, *25*, 101-107.
[5] T. Fujitani, J. Nakamura, *Catal. Lett.* **1998**, *56*, 119-124.
[6] R. Burch, R. J. Chappell, S. E. Golunski, *Catal. Lett.* **1988**, *1*, 439-443.
[7] Y. Kanai, T. Watanabe, T. Fujitani, M. Saito, J. Nakamura, T. Uchijima, *Catal. Lett.* **1994**, *27*, 67-78.
[8] R. G. Herman, K. Klier, G. W. Simmons, B. P. Finn, J. B. Bulko, T. P. Kobylinski, *J. Catal.* **1979**, *56*, 407-429.
[9] J. C. Frost, *Nature* **1988**, *334*, 577-580.
[10] M. Behrens, F. Studt, I. Kasatkin, S. Kühl, M. Hävecker, F. Abild-Pedersen, S. Zander, F. Girgsdies, P. Kurr, B.-L. Kniep, M. Tovar, R. W. Fischer, J. K. Nørskov, R. Schlögl, *Science* **2012**, *336*, 893-897.
[11] M. Kurtz, N. Bauer, C. Buscher, H. Wilmer, O. Hinrichsen, R. Becker, S. Rabe, K. Merz, M. Driess, R. A. Fischer, M. Muhler, *Catal. Lett.* **2004**, *92*, 49-52.
[12] J. D. Grunwaldt, A. M. Molenbroek, N. Y. Topsoe, H. Topsoe, B. S. Clausen, *J. Catal.* **2000**, *194*, 452-460.
[13] R. N. d'Alnoncourt, X. Xia, J. Strunk, E. Löffler, O. Hinrichsen, M. Muhler, *Phys. Chem. Chem. Phys.* **2006**, *8*, 1525-1538.
[14] M. Saito, T. Fujitani, M. Takeuchi, T. Watanabe, *Appl. Catal. A* **1996**, *138*, 311-318.
[15] M. V. Twigg, M. S. Spencer, *Top. Catal.* **2003**, *22*, 191-203.
[16] M. Kurtz, H. Wilmer, T. Genger, O. Hinrichsen, M. Muhler, *Catal. Lett.* **2003**, *86*, 77-80.
[17] M. Behrens, S. Zander, P. Kurr, N. Jacobsen, J. Senker, G. Koch, T. Ressler, R. W. Fischer, R. Schlögl, in *Performance Improvement of Nano-Catalysts by Promoter-Induced Defects in the Support Material: Methanol Synthesis over Cu/ZnO:Al*, submitted to *J. Am. Chem. Soc.*
[18] J. Sloczynski, R. Grabowski, A. Kozlowska, P. Olszewski, M. Lachowska, J. Skrzypek, J. Stoch, *Appl. Catal. A* **2003**, *249*, 129-138.
[19] S. D. Jones, H. E. Hagelin-Weaver, *Appl. Catal. B* **2009**, *90*, 195-204.
[20] J. P. Breen, J. R. H. Ross, *Catal. Today* **1999**, *51*, 521-533.
[21] Y. Nitta, O. Suwata, Y. Ikeda, Y. Okamoto, T. Imanaka, *Catal. Lett.* **1994**, *26*, 345-354.
[22] S. K. Ihm, S. W. Baek, Y. K. Park, J. K. Jeon, *ACS Sym. Ser.* **2003**, *852*, 183-194.
[23] C. Baltes, S. Vukojevic, F. Schüth, *J. Catal.* **2008**, *258*, 334-344.
[24] P. Kurr, I. Kasatkin, F. Girgsdies, A. Trunschke, R. Schlögl, T. Ressler, *Appl. Catal. A* **2008**, *348*, 153-164.
[25] H. Fan, H. Zheng, H. Li, *Front. Chem. Eng. China* **2010**, *4*.
[26] M. Behrens, *J. Catal.* **2009**, *267*, 24-29.
[27] A. Coelho, 4.2 ed., Bruker AXS GmbH, Karlsruhe, Germany, **2003-2009**.
[28] D. A. M. Monti, A. Baiker, *J. Catal.* **1983**, *83*, 323-335.
[29] P. Malet, A. Caballero, *J. Chem. Soc. Faraday T. 1* **1988**, *84*, 2369-2375.
[30] D. G. Barton, M. Shtein, R. D. Wilson, S. L. Soled, E. Iglesia, *J. Phys. Chem. B* **1999**, *103*, 630-640.
[31] B. Ravel, M. Newville, *J. Synchrotron Rad.* **2005**, *12*, 537-541.
[32] M. S. Spencer, *Catal. Lett.* **2000**, *66*, 255-257.

[33] B. Bems, M. Schur, A. Dassenoy, H. Junkes, D. Herein, R. Schlögl, *Chem. Eur. J.* **2003**, *9*, 2039-2052.
[34] P. Porta, S. Derossi, G. Ferraris, M. Lojacono, G. Minelli, G. Moretti, *J. Catal.* **1988**, *109*, 367-377.
[35] S. Zander, B. Seidlhofer, M. Behrens, *submitted to Inorg. Chem.*
[36] M. Behrens, F. Girgsdies, A. Trunschke, R. Schlögl, *Eur. J. Inorg. Chem.* **2009**, 1347-1357.
[37] M. Behrens, F. Girgsdies, *Z. Anorg. Allg. Chem.* **2010**, *636*, 919-927.
[38] G. J. Millar, I. H. Holm, P. J. R. Uwins, J. Drennan, *J. Chem. Soc. Faraday Trans.* **1998**, *94*, 593-600.
[39] L. M. Plyasova, T. M. Yureva, T. A. Kriger, O. V. Makarova, V. I. Zaikovskii, L. P. Soloveva, A. N. Shmakov, *Kinet. Catal.* **1995**, *36*, 425-433.
[40] H. Gomez, M. de la L. Olvera, *Mat. Sci. Eng. B Solid* **2006**, *134*, 20-26.
[41] L. Zhao, G. Shao, S. Song, X. Qin, S. Han, *Rare Metals* **2011**, *30*, 175-182.
[42] M. Zhou, H. Zhu, Y. Jiao, Y. Rao, S. Hark, Y. Liu, L. Peng, Q. Li, *J. Phys. Chem. C* **2009**, *113*, 8945-8947.
[43] J. D. Ye, S. L. Gu, S. M. Zhu, S. M. Liu, Y. D. Zheng, R. Zhang, Y. Shi, *Appl. Phys. Lett.* **2005**, *86*.
[44] N. Roberts, R. P. Wang, A. W. Sleight, W. W. Warren, *Phys. Rev. B* **1998**, *57*, 5734-5741.
[45] H. Matsui, H. Saeki, H. Tabata, T. Kawai, *J. Electrochem. Soc.* **2003**, *150*, G508-G512.
[46] M. M. Günter, B. Bems, R. Schlögl, T. Ressler, *J. Synchrotron Rad.* **2001**, *8*, 619-621.
[47] G. Fierro, M. LoJacono, M. Inversi, P. Porta, F. Cioci, R. Lavecchia, *Appl. Catal. A* **1996**, *137*, 327-348.
[48] M. Behrens, A. Furche, I. Kasatkin, A. Trunschke, W. Busser, M. Muhler, B. Kniep, R. Fischer, R. Schlögl, *Chem. Cat. Chem.* **2010**, *2*, 816-818.

Appendix

Chapter 6: Final Summary and Conclusion

The results presented in this work give insights into preparation of methanol synthesis catalysts by systematic investigation of the precursor chemistry. In particular, they allow to better understand the crucial role of the zincian malachite, $(Cu,Zn)_2(OH)_2CO_3$, precursor phase and its properties on the performance of the resulting catalysts.

Zincian malachite was prepared by co-precipitation, followed by aging, filtrating, washing and drying. Subsequent calcination and reduction led to the catalytically active Cu/ZnO catalyst. Co-precipitation (Cu:Zn = 70:30) was performed in a pH- and temperature-controlled (338 K) manner and enabled homogeneous distribution of the metal ions in the amorphous suspended solid, a Cu,Zn hydroxide carbonate, which transformed into crystalline product during aging. The influence of synthesis parameters, especially in the early stages of catalyst preparation was investigated.

The aging step of Cu,Zn hydroxy carbonates is critical with regard to the incorporation of Zn into zincian malachite and was investigated by *in-situ* energy dispersive x-ray diffraction and *in-situ* UV-Vis spectroscopy. To study the aging process independently, it had to be decoupled from the prior co-precipitation step by a continuous preparation. The obtained "unaged" amorphous precursor phase was transformed to crystalline zincian malachite under controlled conditions by aging in solutions of similar composition to the mother liquor. By varying the pH-values (5.0-8.0), two different aging mechanisms were found. Low pH-values (5.0-6.5) showed direct co-condensation of Cu^{2+} and Zn^{2+}. This mechanism led to higher Zn incorporation as indicated by the shifted position of the $20\bar{1}$ reflection. The second pathway was observed for pH ≥ 7 and showed simultaneous initial crystallization of Cu-rich malachite and a transient Zn-storage phase, sodium zinc carbonate. This intermediate re-dissolved and allowed for enrichment of Zn into malachite at pH ≥ 7 at later stages of aging. As a function of different aging conditions, a variation of the Zn content in zincian malachite between ca. 24 and 29% was observed despite the same nominal Zn-content in the starting material of 30% indicating that a varying fraction of Zn was present in an undetected phase "Zn↓" acting as a sink for Zn. Variation of temperature (at pH 7) only led to gradual changes. Thus, the acidity of the aging medium was identified as the most critical synthesis parameter to determine the final Zn-content in zincian malachite. Interestingly, Zn incorporation was independent of the crystallization mechanism. Even in the absence of Na^+, suppressing the transient crystallization of the sodium zinc carbonate storage phase, a lower degree of Zn incorporation was observed in the final

111

Appendix

sample at pH 7, although the reaction was following the direct co-condensation mechanism. The effect of individual synthesis parameters like temperature or acidity during catalyst preparation can be rationalized on basis of the complex chemistry of precursor aging: They should be optimized to give a low amount of Zn↓ and a maximal Zn-substitution in malachite approaching the nominal Cu:Zn ratio of the synthesis.

Application of different pH-values in the range of pH 6-9 during co-precipitation and aging in a batch synthesis also has a highly reproducible influence on the precursor chemistry. Rietveld refinement was performed on the XRD patterns of the precursors. For pH 6.0, large fractions of the undesired Zn-rich by-phase aurichalcite were found besides zincian malachite as the main phase. Application of pH values ≥ 6.5 led to higher phase fraction of zincian malachite at the expense of aurichalcite with the consequence, that more Zn was introduced into the zincian malachite phase. Samples prepared at pH 7.5 and higher showed a split up signal of the $20\bar{1}$ reflection indicating inhomogeneous distribution of Zn within two different zincian malachite phases. Samples prepared at 6.0 ≤ pH ≤ 7.0 showed a better homogeneity of the Zn distribution. Thus, precursors in this sample series can be characterized by the degree of Zn incorporation into the zincian malachite phase and also the homogeneity of the Zn distribution within this compound. The largest CuO domain sizes were found for calcined samples prepared at pH 6.0. Cu surface areas, which are a prerequisite for the performance of the reduced Cu/ZnO catalysts revealed similar values in the range of 18 to 20 $m^2\ g^{-1}$. Only samples prepared at pH 8.5 showed a larger Cu surface area of around 25 $m^2\ g^{-1}$.

The abovementioned results revealed the complexity of the interplay of synthesis parameters during catalyst preparation by co-precipitation. The properties of the precursor materials obtained by aging of the co-precipitate influence the structural properties, which in turn will affect the performance of the final catalysts. Unfortunately, directly tracking back the catalytic performance to the synthesis pH in a simple synthesis parameter–structure–performance relationship is difficult as variation of the parameter pH induced numerous simultaneous changes in the precursor material that lead to different and partially compensating effects on the resulting catalyst.

ZnO is known to act as a spacer for the single Cu particles in the Cu/ZnO catalyst and to enable Cu-ZnO synergy which beneficially affects the activity. MgO was investigated to act as a substitute for ZnO. The geometric influence turned out to be better compared to ZnO but the synergetic effect of Cu and ZnO during methanol synthesis from $CO_2/CO/H_2$ was unequaled.

Both geometric and synergetic effects were combined by preparation of Cu/MgO/ZnO sample which exhibited a higher activity than Cu/ZnO and Cu/MgO. Changing the feed gas to CO/H_2, Cu/MgO was most active.

Industrial methanol synthesis catalysts are promoted by small amounts of refractory oxide, typically Al_2O_3. In the present thesis, the effect of Ga as a promoter in the Cu,Zn catalytic system for methanol synthesis was investigated by preparing a sample series with increasing Ga concentration. Already small Ga contents up to 3 mol% changed the characteristics of the samples dramatically. Despite the charge mismatch, some Ga^{3+} was incorporated into the zincian malachite precursor phase. Higher Ga loadings led to a Ga segregation and formation of amorphous $ZnGa_2O_4$ spinel in the final catalyst. Thus, the microstructure of the catalyst became inhomogeneous with Ga-rich domains of presumably lower catalytic activity. After calcination, some of the Ga was incorporated in the ZnO which was verified by X-ray absorption near edge structure spectroscopy. After reduction, the Cu surface area was doubled and the methanol synthesis activity increased by 60% compared to the binary Ga free Cu,Zn reference system. The promoting effect was mainly traced back to a geometric contribution: The Cu dispersion was increased by Ga due to better dilution of Cu^{2+} in the precursor by Zn^{3+} and Ga^{3+}, leading to a better interdispersion of metallic and oxidic components and a higher porosity of the resulting Cu/ZnO aggregates. This effect led to an increase of the accessible Cu surface area. Apparently, the intrinsic activity of the exposed Cu surface area was lowered by the presence of Ga. This effect is related to the fraction of Ga species incorporated into the ZnO phase after calcination of the precursor, leading to a modification of the properties of ZnO associated with a negative effect on the well-known Cu-ZnO synergy. This might also be enhanced by less embedment of the Cu particles leading to less interface contact with ZnO. The functionality of Ga promotion depended critically on the homogeneous distribution of Ga. The best distribution was achieved by incorporation into the zincian malachite precursor phase and a linear correlation of the (Zn,Ga) content in this phase with the catalytic activity of the final catalyst was observed.

In summary, these systematic studies on an applied and highly complex catalytic system like Cu/ZnO/X provide a better understanding of the chemistry underlying catalyst preparation and, most important, reveal relationships between synthesis parameters, microstructure and activity that help to explain the role of the structural promoter phase X. These findings shall not only contribute to the fundamental knowledge about this catalyst, but also guide the way to a more rational catalyst design in the future.

i want morebooks!

Buy your books fast and straightforward online - at one of the world's fastest growing online book stores! Environmentally sound due to Print-on-Demand technologies.

Buy your books online at
www.get-morebooks.com

Kaufen Sie Ihre Bücher schnell und unkompliziert online – auf einer der am schnellsten wachsenden Buchhandelsplattformen weltweit!
Dank Print-On-Demand umwelt- und ressourcenschonend produziert.

Bücher schneller online kaufen
www.morebooks.de

OmniScriptum Marketing DEU GmbH
Heinrich-Böcking-Str. 6-8
D - 66121 Saarbrücken
Telefax: +49 681 93 81 567-9

info@omniscriptum.de
www.omniscriptum.de

Printed by Books on Demand GmbH, Norderstedt / Germany